A. U. Mirzaeva
C. X. Umrkulova
F.D. Akramova

Ixoid ticks are ectoparasites of animals in Uzbekistan

A. U. Mirzaeva
C. X. Umrkulova
F.D. Akramova

Ixoid ticks are ectoparasites of animals in Uzbekistan

Imprint

Any brand names and product names mentioned in this book are subject to trademark, brand or patent protection and are trademarks or registered trademarks of their respective holders. The use of brand names, product names, common names, trade names, product descriptions etc. even without a particular marking in this work is in no way to be construed to mean that such names may be regarded as unrestricted in respect of trademark and brand protection legislation and could thus be used by anyone.

Cover image: www.ingimage.com

This book is a translation from the original published under ISBN 978-3-330-33003-0.

Publisher:
Sciencia Scripts
is a trademark of
Dodo Books Indian Ocean Ltd. and OmniScriptum S.R.L publishing group

120 High Road, East Finchley, London, N2 9ED, United Kingdom
Str. Armeneasca 28/1, office 1, Chisinau MD-2012, Republic of Moldova, Europe
Printed at: see last page
ISBN: 978-620-7-85189-8

ACADEMY OF SCIENCES OF THE REPUBLIC OF UZBEKISTAN

Institute of Botany and Zoology

A.U. Mirzaeva, S.H. Umrkulova, F.D. Akramov

IXOID TICKS ECTOPARASITES

ANIMALS OF UZBEKISTAN

Edited by.
Academician of the Academy of Sciences of RUz, Doctor of Biological Sciences,
Professor D.A. Azimov

UDC 576.895.42.

The monograph is devoted to the fauna, peculiarities of ecology and distribution of ixodoid ticks in biogeocenoses of Uzbekistan. Physiological and biochemical mechanisms of tick-host interrelations are analyzed. Toxic characteristics of bioactive substances of salivary glands of dominant species of ixodid and argas ticks are presented. The publication is intended for zoologists, parasitologists, entomologists, veterinarians, workers of sanitary and epidemiological services, students of biological, veterinary and medical profiles.

INTRODUCTION

One of the global threats faced by humanity at the turn of the XX and XXI centuries is the deterioration of environmental conditions, which causes a reduction in the biological diversity of the planet. The conservation of all forms of living organisms has become one of the priority tasks in the overall strategy for the survival of mankind at the global, regional and national levels. In 1995 Uzbekistan ratified the Convention on the Conservation of Biodiversity and took responsibility for the conservation of natural complexes on its territories.[1] At the same time, in many regions of the country the stage of inventory of many groups of invertebrate animals has not yet been completed. This fully applies to ticks, one of the groups of spiders, which include both useful and harmful groups for human activity. The harm of pathogens in global agriculture, estimated at $1.4 trillion[2] . Ixodoidea mites represent a group of highly specialized red-sucking arthropods with two families - Argasidae (Argasinae and Ornithodorinae) and Ixodidae (Ixodinae and Amblyomminae), which are obligate temporary parasites of vertebrates and obligate bloodsuckers, feeding on blood at all phases of development. About 24 species of ixodoid ticks are found in Uzekistan. Exceptional practical importance of ixoid ticks as ectoparasites of farm animals and humans, which are the vectors of many vector-borne diseases, has turned the study of this group into an independent section of parasitology in the development of which, along with zoologists, active participation of specialists in the field of veterinary medicine and general biology. Despite some studies conducted in the past, the fauna of ixodids and argasids and their biology and ecology in Uzbekistan remain poorly studied.

During the years of independence, our country has been carrying out large-scale works on reforming all branches of the agro-industrial complex. Special attention is paid to the preservation of farm animals and increasing their productivity. Based on the program activities, certain results have been

---

[1] National strategy of Uzbekistan on preservation of biological diversity. Tashkent, 1998. - 135 c.
[2] http: // www. fao ozg/ docrep. / 018/ i3300e/ i3300e/pdf

achieved in this direction, including in the livestock sector.

In the process of development of livestock industries, along with other factors, the most noticeable negative impact of pathogens, endo- and ectoparasites, which include numerous species of ixodoid ticks. From this point of view, the problem of tick control is one of the primary objectives of research in the field of parasitology. This involves comprehensive studies of bloodsucking tick communities, pathways and factors of spread, biology and ecology of dominant species using experimental methods of parasitology.

As a result of studies conducted in many regions of the world, a number of scientific results have been obtained. The greatest number of studies is devoted to elucidation of epizootological and epidemiological role of ticks in natural foci and urbanized areas.

At the same time, morphological and biological features of ticks are of exceptional interest for elucidating the processes of circulation in their organism of various groups of pathogens - microbes, viruses, protozoa and other organisms and the mechanism of their transmission to vertebrates, as well as the role of bioactive substances of salivary glands of ticks in the realization of these processes.

Despite certain achievements in the study of ticks in recent years, there are a number of problematic issues that urgently require their solution, both in individual regions and countries. Among these priority areas can be formulated as an explanation of differences in the ability to transmit pathogens of infectious and parasitic diseases of animals and humans; different sensitivity of ticks to acaricides; different toxicity of ticks' diet for hosts and many other practical issues should be sought in the species specificity of physiological and biochemical processes, which in some cases are manifested at the molecular and genetic level. Thus, works in the field of blood-sucking mites should not be limited only to the study of common features characteristic of this group of parasites, but also strive to reveal interspecific differences at the structural and functional levels. These studies are particularly relevant due to the need to address two important practical issues. One of them concerns the development of forecasting of changes in the population size of ixodid and argas ticks, and the other - the methods of possibilities (expansion of their range) of residence of these arthropods when they are dispersed by their feeders within and outside their range. It seems expedient to elucidate the role of different populations in epizootic processes occurring in natural foci and urbanized areas on the basis of quantitative indicators of the relationship in the experiment and knowledge of the ecology of vectors.

CHAPTER I. BIOCENOTIC BASIS FOR THE STUDY OF IXODOIDEA BLOODSUCKING MITES-ECTOPARASITES OF INVERTEBRATES ANIMALS

Mite families Ixodoidea represent a group highly specialized blood-sucking arthropods - temporary obligate ectoparasites of vertebrates, including humans. By general recognition, this group unites representatives of two families - Ixodidae and Argasidae. About 100 species of ixodoid ticks are registered in the vast territory of the countries of the Independent States, and the world fauna includes more than 895 species (Balashov, 1982; Guglielmone et al,2010).

Exceptional practical importance of ixodoid ticks as ectoparasites of agricultural commercial animals, but especially as vectors of pathogens of many infectious and parasitic diseases has attracted the attention of zoologists, parasitologists, entomologists and other specialists in the field of veterinary medicine. Based on the practical needs of public health and veterinary medicine, the ecological features of the most important vectors among Ixodoidea, primarily the faunistic complexes of Ixodidae and Argasidae, their life cycles, the regulation of the latter by environmental factors, the activity of ticks in natural and urbanized biotopes and their relationship with various vertebrate hosts have been studied in detail. A significant number of studies are devoted to the study of morphological, physiological and biochemical features of mites in terms of elucidating their adaptations to parasitism. Such close and diverse relationships of bloodsucking mites with feeders and various pathogens - microorganisms are usually explained by the antiquity of parasitism and the diversity of biocenotic relationships of terrestrial hosts.

The main content of numerous works carried out by researchers of scientific and educational institutions of the world concerns various aspects of participation of bloodsucking ticks, and primarily ixodoid ticks, in the transmission of pathogens of vector-borne diseases of animals and humans and defense against their attack (Balashov, 1982). At the same time, various aspects of their harmfulness as ectoparasites and parasite-host relations are much less well covered.

Argas mites were found to be predominantly nest burrowing parasites, whose habitat is characterized by a relatively stable moisture-temperature regime. Ixodes mites are grazing parasites and, in relatively smaller numbers, nesting burrowing parasites. These and other features in the life of the above groups of ticks have been the subject of study by researchers, in the context of the harmfulness of bloodsuckers to animals and humans. The damage caused by ectoparasites, including diseases transmitted by them, is estimated at 7 billion dollars annually in the world livestock industry, according to FAO (Harrow et al., 1991).

All this served as a basis for critical analysis and generalization of the available literature on ixodoid ticks, and its systematization from the angle of viewpoint on the types of relationships of ectoparasites of terrestrial vertebrates of Uzbekistan.

Studies on fauna, biology, ecology, distribution and importance of the studied groups of ticks as ectoparasites of wild and domestic animals of Uzbekistan (Bernadskaya, 1959; Muratbekov, 1

949-1954;

Muratbekov, 1959; Kuzybaeva, 1961 - 1969; Nadyrov, 1962; Serzhanov, 1964; Muhammedkulov, 1970; Kuklina - 1976) are fragmentary, and the available data are rather outdated, which are of historical interest only.

Other studies on the role of ixodoid ticks in the transmission of infectious and parasitic diseases of birds, mammals, including humans are reflected in the works of a number of authors (Galuso et al., 1959; Rasulov, 2001).

The role of blood-sucking ticks in the spread of very serious diseases of animals and humans in the territory of Uzbekistan and neighboring countries is summarized by U.Y. Uzakov (1974), where it is reported about the participation of species of ixodoid ticks in the transfer of pathogens of viral, bacterial and parasitic diseases.

According to generalized data of the authors (Uzakov, 1974; Kuklina, 1976) 40 species of ixodoid ticks belonging to the families Argasidae and Ixodidae are registered in biogeocenoses of Uzbekistan.

Argaceous ticks are represented by 7 species, ixodes - 33 species, many of which are widely distributed in natural and urbanized areas of the Republic of Karakalpakstan, Fergana and Zarafshan valleys, south of Uzbekistan. As it was established, they can parasitize terrestrial vertebrates - reptiles, birds and mammals.

The 80-90 years of XX century were marked by a new upsurge in the study of harmful activity of bloodsucking mites ectoparasites on the productivity of poultry and livestock sectors of our country. This is evidenced by the results of the study of ixodoid mites of the fauna of Uzbekistan (Raslov et al., 2001; Abdurasulov, 2006; Kazakov, 2010). At the same time, the studied ticks as vectors of pathogens of blood-parasitic diseases of farm animals epizootic situation, practically, in all regions of the country.

Ixodoid ticks united in the families Argasidae and Ixodidae are quite widespread in many regions of the world. They are both ectoparasites and vectors of a number of diseases of vertebrate animals, including humans, causing serious problems for veterinary and public health (Balashov, 1982; Alekseev, Kondrashova, 1985; Gothe, Neitz, 1991; Alekseev, 1993; Hillyard, 1996; Shevkoplyas, 2002; Magometov, 2004; Bogdanova, 2008; Ahmed et al, 2007; Anderson, Magnarelli, 2009, Baker, Craven, 2003; Dennis, Piesman, 2005; Durden et el., 2001; Estrada - Pena et al., 2004, 2006, 2010; Fry et al., 2009; Fukumoto et al.,2006; Jongejan, Uilenberg, 2004; Kaaya, 2003; Kanturk, Yigit, 2011; Kinsey, Durden, 2000; Knipling, Stellman, 2000; Mans et al., 2002; Mbati et al., 2002; Pantaleoni et al., 2010; Permin et al., 2002; Ramamoorthi et al., 2005; Samish, Alekseev, 2001; Shah et al., 2004, 2006;

Spiewak et al., 2006;

Telmadarrehei et al., 2007; Tolleson et al., 2007; Guglielmone et al., 2010). These materials also indicate the scale of dispersal of bloodsucking mite communities in regions of the world and their role in the pathology of humans, agricultural and wild animals, mainly birds and mammals.

The next important direction is devoted to the systematics and taxonomy of ixodoids. There is no unified opinion in this issue. Without going into a detailed analysis of the existing systems of the superfamily Ixodoidea of the world fauna, we find the system of the families Argasidae, Ixodidae and Nuttalliellidae proposed by A. A. Guglielmone et al. (2010) to be the most perfect. The authors provide taxonomies of 896 species: Argasidae - 193, Ixodidae - 702 and Nuttalliellidae - 1, the latter being a monotypic family with a single species, *Nuttalliellidae namaqua*. The authors consider these three families as a part of the Ixodidae order. We adhere to this system.

At present, the problem of parasite-host relations is central in ecological parasitology. The elucidation of quantitative and qualitative aspects of relationships between parasites and their hosts determines the nature of these relationships (Kennedy, 1978; Balashov, 1982; Alekseev and Kondrashova, 1985), which are also characteristic of bloodsucking ticks. According to the character of relationships with vertebrate hosts and habitat types among Ixodoidea, species or groups with pasture-understory and nesting-nesting types of parasitism are distinguished. The nest burrowing type is characteristic of Argas mites; their whole life cycle, including feeding on the host, is carried out inside burrows, nests and other microhabitats inhabited by them. The majority of Ixodidae is characterized by the pasture-undermining type of parasitism (Balashov, 1967).

A specific feature of all developmental phases of ixodid ticks and larvae (rarely nymphs) and some species of argasids is a multi-day feeding on the vertebrate body and absorption of a huge amount of blood and metabolic products during this period. Consequently, ticks lead an ectoparasitic lifestyle, and the vertebrate body serves not only as a source of food, but also as a true habitat for them.

The pathogenic effect of bloodsucking mites on vertebrates is diverse. This is, first of all, the toxic effect of saliva (Balashov, 1982). Under the influence of salivary gland components of a number of species of ixodid and argas ticks, pronounced paralysis and death of vertebrate animals are observed in case of simultaneous parasitization of a large number of ticks. Toxicosis of animals in Central Asian countries is frequently observed when attacked by ticks of the genera *Ixodes, Boophilus, Dermacentor, Argas, Ornithdoros.*

In recent years, components of salivary glands of ixodid and argas ticks have been intensively studied in many scientific centers of the world in order to explain the mechanism of action of toxic substances (Balashov, 1982; Amosova, 2008; Kazakov, 2010; Orlov, Gelashvili, 1985; Andrade et al, 2005; Bergstrom et al., 1999; Bowman, Sauer, 2004; Brossard, Wikel, 2004; Chimikara et al., 2010; Chmelar et al., 2011; Corral - Rodriguez et al., 2009; Donat et al., 1981; Drummond, 1983; Francischetti et al, 2009; Chadeer et al., 2005; Gothe, Neitz, 1991; Kazakov, Abubakirova, 2007; Kazimirova, 2008; Koh et al., 2007; Lagos, Thies, 1969; Lysyk, Majak, Veira, 2005; Mans, Gothe, Neitz, 2002, 2004; Maritz et al, 2001; Maritz - Olivier et al., 2007; Masina, Broady, 1999; Matovu, Olila, 2007; Miller et al., 2005; Motoyashiki, Tu, Azimov et al, 2003; Tu, Motoyashiki, Azimov, 2005; Paveglio et al., 2007; Ramamoorthi et al., 2005; Rezaie, 2004; Ribeiro, Francischetti, 2002; Rolla, 2004; Sauer et al, 2000; Schwan et al., 2003; Simons et al., 2011; Soltys et al., 2009; Stoll et al., 2007; Tomalski et al., 1989; Vandier et al., 2002; Viljaen et al., 1986; Zingali et al, 1993; Spiewak et al., 2006). As a result of the conducted

Physiological, biochemical and toxicological studies of salivary gland secretions of a number of species of ticks Ixodidae and Argasidae have identified bioactive substances - anticoagulants, phospholipases and other components. The structures and molecular masses of anticoagulants and inhibitors have been determined (Motoyashiki et al., 2003; Tu et al., 2005).

The important biological peculiarity of these ticks as vectors of many groups of pathogens of diseases of productive animals dictates complex studies aimed at reducing the population of dominant species or groups of ixodoids and their regulation in natural and urbanized areas.

In recent years, a significant number of works have been published devoted to the development of control methods using chemical, plant preparations, and biological means to regulate the number of bloodsucking mites (Bogdanova, 2008; Podboronov, 1993; Shashina, 2007; Shevkoplyas, 2002; Abdel-Shafy et al, 2007; Abduz Zohir et al., 2009; Al - Rajhy et al., 2003; Anderson, Magnarelli, 2009; Andrew et al., 2004; Cairoll et al., 2004; Chandler, et al., 2000; Choudhary et al, 2004; Chungsamarnyart, Jansawan, 2001; Chungsamarnyart et al., 1994; Dautel, 2004; Dusbabek et al., 1997; Dutra et al., 2004; Fernandes et al., 2008; Foil, 2004; Frazzon et al, 2000; Fernandez - Ruvalcaba et al., 1999; Gindin et al., 2001; Gothe, 1984, 1999; Habib Sewify, 2002, 2010; Hassanain et al., 1997; Hepburn et al., 2007; Hornbostel et al., 2005; Jansawan et al., 1993; Kaaya, 2000, 2003; Kaaya et al, 1995; Kandil et al., 1999; Khater, Ramadan, 2007; Khudrathulla, Jagannath, 1998; Kumar et al., 2000; Lundh et al., 2005; Magano et al., 2008; Martin et al., 2001; Massond et al., 2005; Mawela, 2008; Miller et al., 1995; Montasser et al, 2005; Onofre et al., 2001; Pandita, Ram, 1990; Pereira, Famadas, 2006; Peter et al., 2005; Pirali - Kheirnbadi et al., 2007; Pourseged et al., 2010; Rajput et al., 2006; Ramadan, 2009; Regassa, 2000; Rebeiro et al., 2007; Salum et al, 2002; Samish et al., 2001, 2004; Sewify, Habib, 2010; Tavassoli et al., 2009; Telmadarehei et al., 2007, 2009; Thembo et al., 2010; Tomalski et al., 1989; Williams, 1991; Wutzler, Sauerbrei, 2004; Simons et al., 2011). The results of the conducted studies showed the effectiveness of a number of preparations of chemical and plant origin against some species of argas and ixodid ticks. The role of preparations from fungal strains as acaricides for the control of tick numbers was especially emphasized. The prospect of *entomopathogenic* fungi

Metarhizium anisopliae as an alternative to chemical ones has been noted (Pourseyed et al., 2010). A number of works report the possibility of using bacterial preparations against animal ectoparasites (Hassanain et al., 1997; Kaaya, 2003).

When studying publications on ixodoid ticks from different regions of the world, the following facts draw attention: ixodid and argas ticks are widely distributed as ectoparasites of terrestrial vertebrates and are involved in the transmission of a number of infectious and parasitic diseases of farm and wild animals, including humans.

All studied mites are obligate temporary parasites, part of their life cycle takes place on the host body, and part - in the external environment. The relationship between the free-living and parasitic phases of biology is determined by the nature of relationships with vertebrate feeders.

Different conditions of existence of free-living and parasitic phases of development of ixoid ticks within one morphological phase contributed to the emergence of deep morpho-physiological and biochemical differences between them. The perfection of adaptations of ixodid and argas ticks to temporary parasitism on terrestrial vertebrates ensured maximum use of animal blood as a food source, providing absorption of a large mass of blood for growth and development. In both groups of mites, the salivary glands produce bioactive substances such as anticoagulants that prevent clotting of blood in the parasite and damaged blood vessels in the host's skin thickness. Components of bioactive substances contained in tick saliva are intensively studied by many researchers in America, Europe and Asia.

Analysis of various works on the control of bloodsucking mites has shown that among the tested chemical, herbal and biological preparations, acaricidal agents have been identified. Among them there are very promising agents, which require further research.

In general, it can be noted that research on ixodoid ticks is carried out in many scientific and educational institutions of the world. Fundamental and applied results have been obtained in the framework of our research concerning ticks of the families Ixodidae and Argasidae of the fauna of Uzbekistan.

1.1 Materials and methods of research

The material was collected during (2008-2016) from biotopes located in the North-Eastern, Central and Southern regions, covering 8 provinces of Uzbekistan (Tashkent, Syrdarya, Jizzak, Samarkand, Bukhara, Navoi, Surkhandarya and Kashkadarya).

Material was collected in all seasons of the year: spring, summer, fall and winter from mite habitats (animal quarters near water bodies, burrows, caves and living quarters) according to known methods (Pospelova - Strom, 1953; Agrinsky, 1962). 136126 specimens of ticks were collected and examined. The dynamics of tick infestation of farm animals (*Bos taurus* dom., *Ovis aries* dom., *Capra hircus* dom., *Equus caballus dom.*, *Camelus dromedarius* dom., *Suis scrofa* dom.) by ixodids and birds (*Gallus gallus* dom., *Meleagris gallopavo* dom.) by argasids was studied by collecting ticks once a month with repetitions from 2008 to 2016. At the same time, some species of wild animals (reptiles, birds and mammals) were periodically examined for tick infestation.

The anticoagulant effect of tick saliva secretion was studied according to known methods (Fry et al., 2009). The methods used in our experiments are also given in the relevant chapters and sections of this work.

CHAPTER II. FAUNA AND ECOLOGY OF IXODOIDEA - BLOODSUCKING MITES OF VERTEBRATES OF UZBEKISTAN

As a result of this research, 24 species belonging to 9 genera and two families, Ixodidae and Argasidae, were found (Fig. 1).

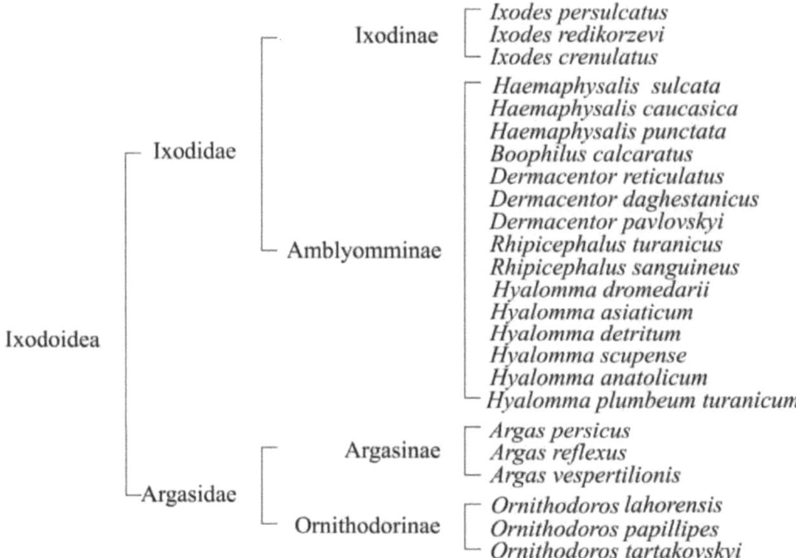

Fig. 1.Taxonomic composition and species diversity of ixodoid ticks in Uzbekistan

The family Ixodidae is represented by 18 species, 17 of them were registered on flat territories, 13- in foothill areas and 10 in mountainous. Representatives of the genus *Hyalomma* were dominant in the plain territories. This genus in our collections constitutes the main background of the ixodid tick fauna (62.6 %) and is represented by 6 species - *H. asiaticum* Sch. et Schlottke, 1929, *H. anatolicum* Koch., 1844, *H. detritum* Sch., 1919, H. *dromedarii* Koch., 1844, *H. scupense* Sch., 1918 and *H. plumbeum* Panzer, 1795.

The most frequent parasites of farm animals are *H. asiaticum* (33.7%), *H. detritum* (22.6%), and *H. anatolicum* (about 20.0%). These species were recorded practically in all regions of Uzbekistan. The highest abundance is observed in the Southern and Central regions. Plains with different climatic conditions are also favorable for *Boophilus calcaratus* Birula, 1895 (16.5%), *Rhipicephalus turanicus* Pomerantzev, 1940 (10. 7%) and *Dermacentor reticulatus* Fabricius, 1794   (2.   2%). Here

some species of genera *Ixodes* and *Haemaphysalis are* also widespread. Representatives of the genera *Ixodes* (18%) were recorded in the foothill areas, *Boophilus* (15.6 %), *Rhipicephalus* (15.1 %) and *Hyalomma* (18.9 %).

In the mountain belt 10 species of genera *Ixodes* (1.4 %), *Boophilus (32.6 %), Rhipicephalus* (4.2 %) and *Hyalomma* (37.0 %) were recorded. The genus *Ixodes* is found mainly in foothill and mountainous zones. Representatives of the genera *Haemaphysalis* and *Dermacentor* are confined to plains. Species of genera *Boophilus, Rhipicephalus* and *Hyalomma have* adapted to all landscape zones.

The family Argasidae in our collections is represented by two genera, *Argas* Latreille, 1796, *Ornithodoros* Koch., 1844.

The genus *Argas is* represented by three species - *Argas persicus* Oken, 1818, *A. reflexus* Fabricius, 1794 and *A. vespertilionis* (Latreille, 1802). *A. persicus is* most widespread both on the plains and in the foothills, where infestation of domestic chickens reaches 35 to 70%. The intensity of infestation fluctuated 7   -147 exv. The genus *Ornithodoros is* represented by three species - *O. papillipes* (Birula, 1895), *O. tartakovskyi* Olenev, 1931 and *O. lahorensis* (Neumann, 1908). They have been recorded both on animals and in farm premises as well as human dwellings. The first species is confined to foothill and mountainous areas and parasitizes cattle, sheep, goats, horses and humans. *O. tartakovskyi is* registered on farm animals in desert and semi-desert landscapes. It is distributed in the valleys of the Amu Darya, Syr Darya and Zarafshan rivers. In natural conditions they are found in burrows of insectivores, rodents and reptiles. The third species, *O. lahorensis, is* found on plains and foothills, most often on sheep (15.5%), with the intensity of infestations ranging from 9 to 85 specimens.

The role of feeding animals in the dispersal of many species of ixodes and argas ticks in natural and urbanized areas of Uzbekistan should be noted. Agricultural and wild animals, migrating, very effectively participate in the dispersal of ticks.

The activity of dominant species of ixodids and argasids depends on seasons of the year and landscape. The activity in the plains is observed from the third decade of February and early March, in the foothill belt - in and April, and in the mountain belt - in late April and early May. The parasitization season of ticks in different zones differs in the timing of their infestation of animals. The maximum infestation of animals, as a rule, is observed in summer. High infestation of cattle with *H. asiaticum* reached 32.5 %. Chickens were infested with *A. persicus* by 29.3 %. Decrease of animal infestation is observed in all zones in fall and significantly in winter.

It follows from the results of our research that in the territory of Uzbekistan ixodoid ticks are currently represented by 24 species. Of these, ixodids comprise 18 species, and argasids - 6. In previous studies (Uzakov, 1972; Kuklina, 1976) 40 species of ixoid ticks were found in domestic and wild animals: Ixodidae - 33 species and Argasidae - 7 species. It should be specified that the majority of species were noted from single findings, single specimens or immature phases. By now, the data of previous studies are noticeably outdated, which is confirmed by recent studies of the acarifauna of the region (Rasulov et al., 2003., Abdurasulov, 2006; Umrkulova et al., 2016). According to the results of studies in recent years, there is a noticeable impoverishment of the ixodoid tick fauna in Uzbekistan. A significant number of species were absent in our collections: *Ixodes redikorzevi emberizae, Haemaphysalis numidiana turanica, H. pavlovskyi, H. concinna, Dermacentor marginatus, D. silvarum, D. nuttulova et al. silvarum, D. nuttalli, Rhipicephalus bursa, R. rossicus, R. pumilio, R. leporis, R. schulzei, Hyalomma aegyptium, H. anatolicum excavatum, H. plumbeum impressum* and *Ornithodoros cholodkovskyi.* In our opinion, the main factor limiting the species composition of ixodoid ticks on the territory of Uzbekistan is human economic activity - large-scale development of natural areas, which contributes to changes in the vegetation cover and habitats of wild animals - feeders of ticks.

In the landscape distribution of the studied species, a special place is occupied by plain territories. Ticks are found practically in all areas with the prevailing agricultural type of development. Agricultural mammals and birds, whose abundance is high, play an important role in maintaining high numbers of a number of species.

The complex of species comprising the fauna of ixodoid ticks in Uzbekistan at the present stage implies monitoring of their numbers in order to improve methods of their control in specific areas.

2.1 Blood-sucking mites of the family Argasidae

As indicated above, we found six species of argas mites belonging to 2 genera and 2 subfamilies in the biogeocenoses of Uzbekistan:

Family Argasidae Canestrini; 1890

Subfamily Argasinae Posp. - Sht., 1946

Genus *Argas* Latreille, 1796

Species: *Argas* persicus Oken, 1818

Argas reflexus Fabricius, 1794

Argas vespertilionis (Latreille, 1802)

Subfamily Ornithodorinae Posp. - Sht., 1946

Genus *Ornithodoros* Koch., 1844

Species: *Ornithodoros papillipes* Birula, 1895

Ornithodoros tartakovskyi Olenev, 1931

Ornithodoros cholodkovskyi Pavl., 1930

Ornithodoros lahorensis (Neumann, 1908)

The detected species of Argas mites are obligate hematophagous and practically have a ubiquitous distribution. As indicated above, the genus *Argas* is represented by three species - *A. persicus, A. reflexus* and *A. vespertilionis.* These species can parasitize both birds and mammals. The most widespread populations of *A. persicus are* in the zone of plains and foothills, where infestation of domestic chickens reaches from 35 to 70%. The intensity of infestation ranged from 7- 147 exv. Among these species, *A. persicus* occupies a dominant position as an ectoparasite of birds.

The genus *Ornithodoros* in our collections is represented by three species - *O. papillipes, O. tartakovskyi* and *O. cholodkovskyi.* They were observed both on animals and livestock and poultry premises. Populations of these species are evenly distributed in Kashkadarya and Surkhandarya provinces. *O. papillipes* and *O. tartakovskyi* are numerous.

O. lahorensis, which occurs more frequently in sheep in all studied regions of the republic, causes a serious disease - tick paralysis. A significant number of ticks was observed in premises used for sheep wintering, less in places of concentration of cattle and horses. Cat ticks are still a very serious ectoparasite of sheep and other animals.

Peculiarities of distribution and ecology of argas mites. All argas mites are obligate temporary hematophagous mites. Feeding on vertebrate blood is a necessary condition without which normal development of mites and their reproduction are impossible (Ginetsinskaya and Dobrovolsky, 1978). Part of the life cycle of argas mites takes place on the body of the vertebrate host, and part in the external environment. The ratio between the free-living and parasitic phases of the life cycle is determined by the nature of relationships with vertebrate feeders. Different conditions of existence of the population of free-living and parasitic stages in the biology of argas mites within one morphological phase caused the emergence of deep morpho-functional differences between them. The main difference of argasid mites in the parasitic phase by the nature of feeding is in repeated blood sucking at the same phase of the life cycle. Argas mites are predominantly nest-breeding ectoparasites, whose habitat is characterized by a relatively stable moisture-temperature regime. All this contributes to the lengthening of the active life period and attacks on animals - feeders.

Of the morpho-functional features of *Argas* mites described in detail by Balashov (1967), the following are of most interest (on the example of the genus *Argas*).

Large mites, body length (hungry) reaches 3 - 12 mm, width 6-8 mm. In general, reproduction of the studied mites is in direct dependence on blood feeding. As for many hematophagous mites, Argasidae are also characterized by the presence of a gonotrophic cycle. This leads to the dependence of the egg maturation process on the processes of blood digestion and assimilation. Absorption of a certain portion of blood by female Argas mites ensures the formation, development and laying of numerous eggs. New portions of oviposition are possible only after repeated blood sucking (feeding). The period running from one act of blood sucking to the next, between which the mites reproduce, is called the gonotrophic cycle. Females of Argasidae undergo up to 4-6 gonotrophic cycles during their life, which contribute to the increase in the number of mites (Fig. 2).

Argas mites are widely distributed in the studied territories of Uzbekistan. They are noted as ectoparasites of birds and mammals inhabiting plains, foothills and mountains (Table 1).

The investigated representatives of genera, occur mainly on the plain, where domestic animals - feeders of argas mites are concentrated.

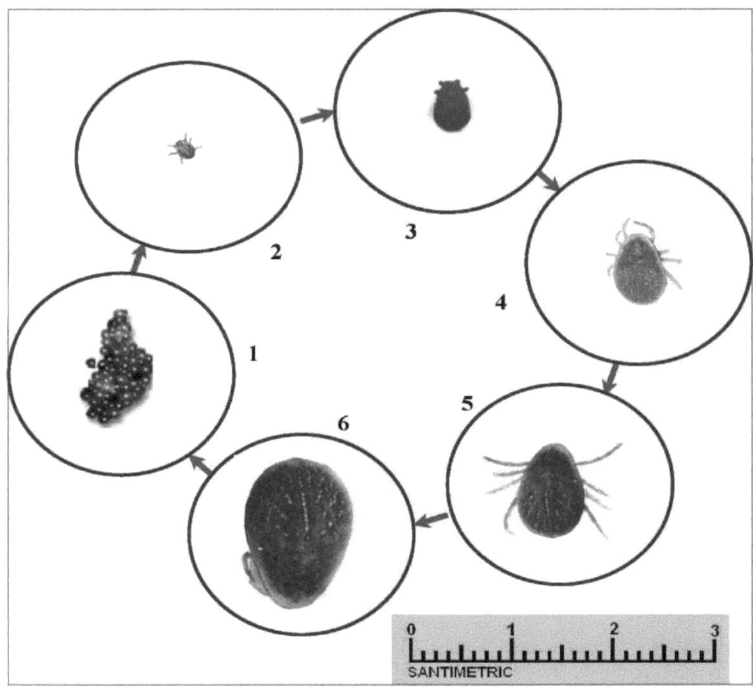

Figure 2. Developmental phases of the mite *Argas persicus*.
1 - eggs of *A. persicus,* 2 - larva, 3 - I nymphal stage, 4 - II nymphal stage, 5 - III nymphal stage, 6 - imago.

Table 1.

Distribution of Argasidae mites in landscape elements

Mite genus	Landscapes		
	plains	foothills	mountains
Argas	+++	+++	+
Ornithodoros	+++	++	-

Note: +++ mass; ++multiple; +small; - absent.

The role of animal feeders in the dispersal of many species of Argas mites in natural and urbanized areas of Uzbekistan should also be noted. Domestic and wild animals infected with ectoparasites, moving (migration) from one territory to another, very effectively participate in the dispersal of the studied mites. In this regard, most species of ticks are registered as ectoparasites of animals inhabiting plain, foothill and mountain zones.

The activity of dominant species of ixodids and argasids depends on seasons of the year and landscape. The activity of ticks in the plain is observed from the third decade of February and early March, in the foothill belt - in March-April and in the mountain belt - in late April and early May. The parasitization season of ticks in different zones differs in terms of their infestation of animals. Tick infestation of animals by certain species (groups) of the studied ticks is correlated with the seasons of the year (Table 2).

Table 2

Dynamics of Argas tick infestation in animals of Uzbekistan depending on seasons of the year

View	Inoculum, %			
	Spring	Summer	Autumn	Winter
Argas persicus	4.1	29.3	10.1	5.6
Ornithodoros tartakovskyi	4.4	27.3	9.5	5.8
Ornithodoros lahorensis	3.1	24.6	8.2	4.2

The maximum infestation of animals, as a rule, is observed in summer. High infestation of sheep with *O. lahorensis* reached 24.6%. Domestic chickens were infested with *Argas persicus* - 29.3%. Decrease of animal infestation is observed in all zones in fall and significantly in winter.

Specificity of parasites in the choice of hosts can be conditioned both by the belonging of the latter to certain taxonomic groups of animals (phylogenetic specificity) and by ecological factors, when the parasite can live on unrelated species of hosts occupying similar ecological niches (ecological specificity)

(Balashov, 1982). In this connection, we conducted a study to find out numerical ratios between three species of the genus *Argas* in urbanized areas of Uzbekistan (Fig. 3).

Collected from 2846 domestic birds and pigeons from poultry farms and pigeon breeders of Tashkent, Surkhandarya and Kashkadarya oblasts, out of 15568 specimens of ticks were representatives of *A.persicus* - 13638 specimens, *A.reflexus* - 1510 and *A.vespertilionis* - 420. Undoubtedly, *A.persicus* (87.6%), *A.reflexus* (9.69%), *A.vespertilionis* (2.69%) dominate the total number of ticks of this genus. Interesting is the fact of significant distribution of the last two species in domestic chickens and turkeys, which in natural conditions are specific parasites of pigeons and bats, respectively. From the presented materials it is possible to make

Fig. 3. Numerical ratio of mite species of the genus *Argas* - parasites of birds in
Uzbekistan

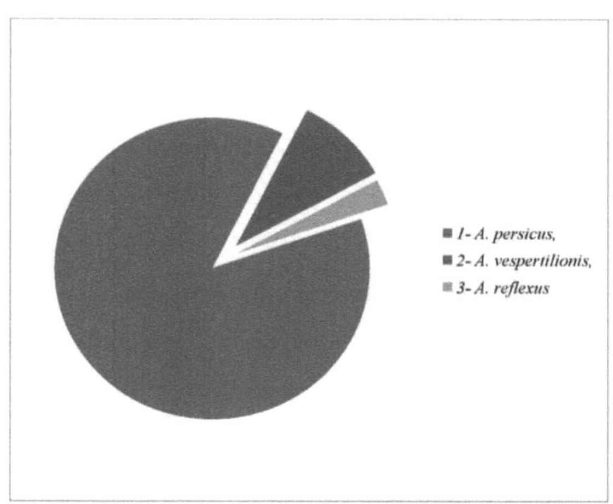

conclusion that mites of *A.reflexus* and *A.vespertilionis* species under appropriate conditions can parasitize other animal species with the resulting consequences.

2.2 Study of gas exchange of mites of the family Argasidae - ectoparasites of birds

Respiration of mites is carried out by the tracheal system, the cuticular lining of which is replaced during molting. The tracheal system in a number of species of the family Argasidae. Argasidae (*Ornothodoros lahorensis, Argas persicus, A. vulgaris, A. reflexus, Ornothodoros coniceps, O. capensis*) is sufficiently developed (Balashov, 1967). Eight main trunks branch from the common tracheal trunk, giving secondary branching in all parts of the body. Branching repeatedly, each tracheal trunk ends with a bundle of the finest tracheoles that braid the internal organs of the mite in an empty network. The rhythm of respiratory movements and intensity of tracheal ventilation depends on the species, its condition and spring conditions.

The present work is devoted to the study of gas exchange in a number of species of mites of the family Argasidae under different temperature conditions, which is a prerequisite for understanding their high viability to the effects of unfavorable environmental conditions.

Argas mites *A.persicus, A.reflexus* and *A.vespertilionis* collected from poultry farms in the southern regions of Uzbekistan were used. Adult (imago) mites were selected for experiments and kept at 20-22^0 C. Measurement of gas exchange (O consumption$_2$ and

CO_2 emission) in insects was performed using a modified Winterstein microrespirometer (Kogan and Shitov, 1967).

A microspirometer connected to a vessel containing an O_2 absorbent (KOH) was used. - (KOH). This reservoir with the absorbent in turn,

is connected to the measuring capillary. The displacement of the solution column in this capillary was used to judge the intensity of gas exchange in mites. In experiments with different temperatures, we used a water thermostat. A microrespirometer chamber with mites was immersed in the thermostat. After one hour, the microrespirometer was removed from the thermostat and connected to the absorption chamber, which had a measuring capillary to assess the intensity of respiration of mites. To quantitatively characterize the gas exchange of ticks, the intensity of O2 consumption and CO2 release in mm^3 per hour per g of individuals was determined under different temperature conditions (10^0 C, 20^0 C, 30^0 C). The intensity of gas exchange in different species differs significantly (Table 3). The materials of the table show that a higher level of gas exchange is characteristic of *A.reflexus* and *A.vespertilionis* mites. *A.vespertilionis, on* the contrary, has a low level of gas exchange, and the differences found are determined by biological features of mites, since gas exchange may represent a single metabolic basis that contributes to the survival of mites in extreme environmental conditions.

Table 3

Level of gas exchange in mites of the family Argasidae at different ambient temperatures

Temperature in 0C	*Argas persicus*		*Argas reflexus*		*Argas. vespertilionis*	
	Por. O$_2$ in mm^3 per g weight per hour	Highlight. CO$_2$ in mm^3 per g of weight per hour	Por. O$_2$ in mm^3 per g of weight per hour	Highlight. CO$_2$ in mm^3 per g of weight per hour	Por. O$_2$ in mm^3 per g of weight per hour	Highlight. CO$_2$ in mm^3 per g of weight per hour
30	298.20±15.4 2.52	217.40±12.11 12.49 0.73	470.40±23.4 1.52	324.40±15.2 1.46 0.68	167.20±11.4 1.89	119.20±10.2 1.75 0.71
20	118.30±11.7	87.20 0.73	268.90± 17.1	221.60 0.82	88.10±9.9	67.80 0.76
10	15.50±10.1 7.6	12.31±5.8 7.2 0.79	28.51±6.3 9.4	25.15±5.6 8.8 0.88	8.80±2.4 10.4	7.60±1.4 8.9 0.86

Note: T-ra - temperature, Consumption - consumption, Emission. - allocation, oblique figures - degree of change (in times) of gas exchange from the control level (control -20^0 C). Control data in the table are marked with a bold line. The small square boxes show the respiratory quotient value.

It is known that abiotic factors significantly affect the vital activity of mites, one of which is the external temperature, which varies greatly by seasons of the year and during the day. Taking into account this fact, we performed a series of experiments to determine gas exchange in mites of the Argasidae family at different ambient temperature conditions (10^0 C, 20^0 C, 30^0 C). Although the vital activity of mites occurs at different environmental temperatures, however, the average temperature for them can be approximately 20^0 C. In this regard, the data obtained by us in laboratory conditions at this temperature, are taken as a control. The data presented in (Table 3) show that lowering the temperature causes a decrease in the intensity of gas exchange in mites, which is accompanied by a decrease in the consumption of O_2 and CO_2 release.

All species of the studied mites show the same regularity in changing the level of gas exchange indices, although there is a certain difference between them. It can also be seen that mites with an initially higher level of metabolism retain a relatively increased level of gas exchange when the external temperature changes.

The mites *A. reflexus* and *A. vespertilionis* turned out to be especially sensitive to temperature effects, in which at decreasing of the medium temperature from 20^0 C to 10^0 C their gas exchange is suppressed approximately 10 times, and in *A. reflexus* such decrease reaches up to seven times. However, when the temperature of the medium rises from 20^0 C to 30^0 C, gas exchange increases only within two times in all studied mite species.

Consequently, comparative consideration of the results obtained on gas exchange of different mite species shows that this parameter of metabolism is very sensitive to temperature influence and is capable of a significant decrease under decreasing temperature conditions.

We also determined the respiratory quotient in mites under different temperature conditions. This index is calculated as a quotient of the division of the volume of CO excreted$_2$ to the volume of O absorbed$_2$. Knowing that this coefficient has a value up to unity in the oxidation of carbohydrates in the insect organism, we obtained values of this index ranging from 0.7 to 0.9 in our experiments.

These values indicate that not only carbohydrates, but also other substances, in particular, proteins, are utilized in the organism of mites. The respiratory quotient of *A. reflexus* was more labile, with a low level of this index at 30^0 C, while at lower temperatures it increased to 0.8 - 0.9. It can be said that the increase in temperature in this species of mites contributes more to the increase in utilization of carbohydrate substances by them. The differences in gas exchange, apparently, are determined by ecological and biological peculiarities of the mites studied by us and may be the metabolic basis of their resistance to unfavorable temperatures. This is also evidenced by a decrease in gas-oxygen exchange in other mite species under extreme environmental conditions (Belozerov, 1966; Malygin and Alekseev, 1989).

In general, our results indicate the lability of gas exchange in mites, which may have an important biological significance when the ambient temperature decreases during the hibernation period, when the latter are subjected to prolonged starvation. A sharp suppression of metabolic processes is probably one of the important factors ensuring the survival of mites in extreme environmental conditions (Mirzaeva et al., 2011).

CHAPTER III. BIOACTIVE SUBSTANCES OF THE SALIVARY

OF THE GLANDS OF ARGAS MITES

It is well known that the composition of saliva of bloodsucking ticks consists of prostaglandins, vasodilators, antiplatelets, immunomodulators and anticoagulation substances associated with their main adaptive mechanisms. Examination of the structural and functional features of saliva revealed the presence of poisonous components. Available anticoagulant substances, prevent the host blood from clotting (Motoyashiki et al.,2003; Tu et al., 2005). Most of the biologically active substances included in tick saliva have a wide range of actions.

Biologically active substances isolated from saliva of a number of mite species possessing anticoagulant and antiplatelet mechanism of action are poorly studied. However, studies of physiological mechanisms of action of the secretion of mites Argasidae and Ixodidae are fragmentary and completely insufficient.

3.1. Anticoagulant and hypotensive effects of salivary gland secretion of *Ornithodoros tartakovskyi* ticks

In the conducted studies on the anticoagulant effect of salivary gland secretion of *O. tartakovskyi* ticks, *the* slowing down of the clotting process of rat and ram blood plasma was noted. Parameters of coagulation were dependent on the concentration of the tested secretions (25 - 150 µg/mL). At the concentration equal to 150 µg/mL, the degree of blood plasma coagulation (compared to the control) decreased to 79.7±5.6% (Fig. 4).

To elucidate the anticoagulant mechanisms of blood plasma at deficiency of X, XI, IX and VIII factors, WHAT (Time of partial activation of thrombin) was used. In this case, the aggregation process of blood platelets is prolonged to a large extent. Figure 3 shows the aggregation process of rat blood platelets. The noted slowing down of anticoagulant process by biologically active substances of mite secretions resulted in 27

attenuation of platelet activity and inhibition of X factors. Some anticoagulant components led to a reduction in platelet platelet formation (Waxman, 1993), others to activation of Xa factors inhibition (Waxman et al.,   1990).   The effect of the complex has been determined

prothrombinase or Xa complex, and thrombin binding during anticoagulant effects (Corral, 2009).

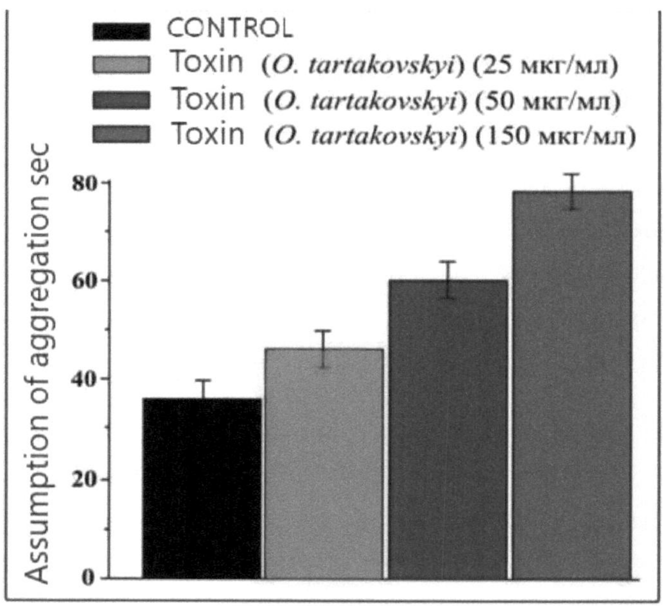

Figure 4. Effect of salivary gland secretion of ticks *Ornithodoros tartakovskyi,*
of different concentrations on the aggregation of rat blood platelets

The components isolated from salivary glands of mites *Ornithodoros moubata* and *O. savignyi, Moubatin* and *Savignin* slowing down blood coagulation processes led to a decrease in thrombin activation, and have an anti-aggregation feature (platelet aggregation) in the blood system. And also monobin isolated from salivary glands of *Argas monolakensis* has anticoagulant effect (Waxman, 1990).

It was found that TAP (tick anticoagulant peptide) protein molecules isolated from salivary glands slow down the activity of FXa factors action on human plasma blood coagulation processes. As a consequence, they revealed the formation of active bonds of TAR protein FXa factor and achieved positive results in the process of their antithrombotic effect in the blood system (Rezaie et al., 2004; Fukumoto et al., 2006).

The results of our studies and literature data show that *O. tartakovskyi* secretions (25 - 150 µg/mL) significantly affect the activity of hemostasis and to a lesser extent the duration of anticoagulation features of blood plasma according to the test (HFAT). Secretions of *O. tartakovskyi,* exert retarding actions of factors X, XI, IX, VII in the circulatory system. Relaxant (relaxing) effect of *O. tartakovskyi* salivary gland components on the contractile activity of a muscle preparation prepared from rat aorta was observed. In turn, the effect of KCl at a venom concentration of 150 µg/mL resulted in regression of the force of contraction of the muscle preparation compared to the control. The half-maximal effect of EC concentration$_{50}$ was 34.7 µg/mL with pD2 (-log EC50) equal to 4.73.

Ions [Ca^{2+}] play an important role in the regulation of contractile activity of smooth muscle cells of blood vessels, and this process is directly related to [Ca^{2+}]i. The following channels are involved in the regulation of this process: channels located in the plasmolemma Ca^{2+} L, as well as activating receptor channels influenced by sarcoplasmic reticulum inositol 1, 4, 5 - triphosphate (IP3R), calmodulin protein (Sanders, 2001), Ca^{2+} -ATPase, located in the plasmolemma Ca^{2+} -ATPase and Na^+ /Ca^{2+} - exchanger and protein kinase linkage.

In order to ensure blood flow in the body of ticks, their salivary glands contain a number of vasodilators - substances that ensure the expansion of blood vessel walls (Rieiro, 2003; Andrade et, al., 2005). Prostaglandins, which have vasodilator properties, have been identified in the salivary glands of ticks of the families Ixodidae and Argasidae. In addition, they have a relaxing effect on smooth muscles of blood vessels and slow down platelet aggregation (Andrade et, al., 2005).

When blood vessel walls are damaged in the process of platelet activation, there is an increase in Ca^{2+} ions associated with a reduction in the activity of smooth muscles of blood vessels. In this connection, the protein molecules calreticulin, available in the salivary glands of ticks, bind Ca^{2+} ions and limit the contraction of smooth muscles of blood vessels (Bowman, 1996).

From the analysis of the results of our studies and literature data it is seen that the action of salivary gland secretions of *O. tartakovskyi* mites on the contractile activity of the rat aorta preparation, blood vessels smooth muscles has a relaxing effect. In turn, it is realized by the action of prostaglandins as a vasodilator and binding action of calreticulin protein molecules with Ca ions^{2+}

The obtained results of research allow that bioactive substances of salivary gland secretions of mites deserve to be used as natural components for creation and improvement of means for correction of vascular pathology.

3.2. Anticoagulants of *Argas persicus* ticks, isolation and testing of their activity

After lyophilic drying, the biological material (152 mg) was dissolved in 4 ml of 0.01 M ammonium hydrogen carbonate (NH_4HCO_3) buffer. The resulting solution was centrifuged at 6000 rpm for 15 minutes. Subsequently, the resulting precipitate. was run on a TSK HW-55f sorbent column (1.6x125 cm) in 0.01 M ammonium hydrogen carbonate (NH_4HCO_3) buffer. The flow rate of buffer through the column was 60 mL per hour. The yield of the required fraction in the eluate was determined spectrophotometrically (UB detector - Uvicord S" -LKB" Sweden) with a wavelength of 280 nm and at 0.1 sensitivity.

The chromotographic analysis was carried out using common methods (Stefanova, 2002). The method of Lowry et al. was also used to determine the protein content of the fractions (Baluda et al., 1980). (Baluda et al., 1980).

Column gel filtration of *A. persicus* lyophilate yielded 5 protein fractions (Fig. 5).

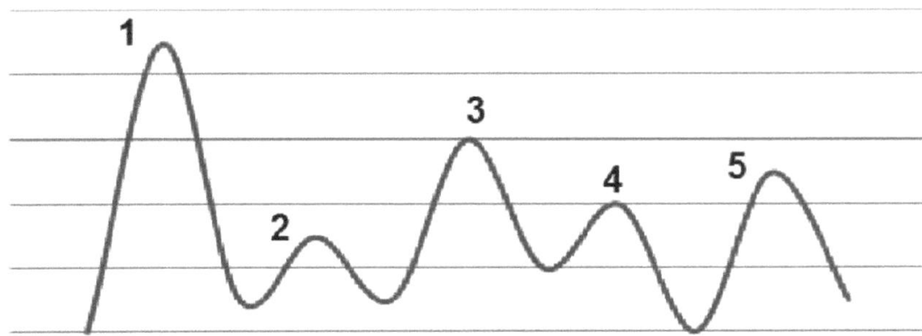

Figure .5. Yield of *Argaspersicus* protein fractions through the column withTSK sorbent
HW-55f (column: 1.6X125CM) in column gel filtration process

It can be seen that the protein fractions shown in the diagram have different heights, that the amount of protein is different in these fractions. The highest protein content occurs in the first and third fractions. We used these data in the subsequent analysis of anticoagulant activity of different fractions.

Determination of amino acid composition of proteins in *A. persicus* phenylthiocarbomyl (PTC) extract by a sensitive method. A 1 mg suspension of salivary gland extract was dissolved in 1 ml of distilled water. To precipitate proteins and polypeptides, 10% perchloric acid was added to this solution and centrifuged at 8000 rpm. Identification of FTC amino acids was performed on a chromatograph (Agilent Nechnologies 1200). The results of analyzing the quantitative content of amino acids in the salivary gland extract of *A. persicus* show (Table 4) that glutamic acid, glutamine, and proline are detected in higher amounts. Somewhat lower levels of tyrosine and
histidine, while other amino acids are found in even smaller amounts.

Table 4

Results of analysis of quantitative amino acid content in

salivary gland extract of *A. persicus*

salivary glands

Amino acid names	Amino acid content (µg/g)
Asparagine acid	0,795858
Glutamic acid	3,010659
Cerin	0,44964
Glycine	0,376835
Asparagine	0,392079
Glutamine	8,546236
Cysteine	0
Threonine	11,9802
Argenin	1,549021
Alanine	2,231724
Proline	4,103328
Tyrosine	4,840961
Valine	1,834037
Methionine	2,017019
Isoleucine	0,717403
Leucine	0,81513
Histidine	5,664841
Tryptophan	1,102432
Phenylalanine	0,472549
Lysine HCl	1,083174
Total:	**51,98312**

The results obtained suggest that such amino acids as glutamic acid, glutamine, and tyrosine play a major role in the protein fractions of *A. persicus* salivary gland extracts. This finding is of some interest for further studies on the structure and composition of anticoagulants.

Based on the obtained data, we carried out studies to determine the calibration activity of *A. persicus* substances with a standard analog (heparin) in in vitro experiments, where the blood clotting time, activated partial thromboplastin time were taken into account.

Typically, studies are performed using a calibration curve for low anticoagulant levels of 0.02-0.10 units and high levels of 0.1-0.6 units. The concentration of anticoagulant is selected so that the enzyme inactivation is linear. The analyte and analog are added to platelet-poor citrate plasma. For this purpose, blood is centrifuged for 20 minutes at 2000 rpm. Blood for the study was obtained dropwise from a mouse ear vein.

In our experiments, we worked with high and low anticoagulant levels of 0.1-0.6 U and 0.02-0.06 U.

The clotting time of whole blood was determined on a water bath using a stopwatch, which was switched on immediately after the beginning of blood flow from the mouse ear into a tube up to 1 ml, to which 0.1 ml anticoagulant solutions were added beforehand.

The obtained data are shown in Tables 5 - 9. As can be seen from the data shown in Table 2, blood clotting time increases with the addition of all drug samples. The greatest effect was observed at a concentration of 0.06 U/mL for heparin and salivary gland samples of *Argas persicus* ticks twice as much as for the control, to which 0.9% saline solution was added.

As can be seen from the data in Table 6 activated partial thromboplastin time, which depends on the activity of factors VIII, IX, XI, XII and prekallikrein involved in the internal mechanism of blood coagulation increased under the influence of heparin and salivary gland samples of *Argas persicus* ticks.

Table 5.

Effect of anticoagulant samples on clotting time at different concentrations on mouse blood in in vitro experiments

grouping	The drug T	Anticoagulant concentration, units/time in min					
		control	0,02	0,03	0,04	0,05	0,06

0	heparin	0,6±0,2	2,5±1,2	4±1,1	4±1,2	5±1,4	6±1,3
1	fraction	-	-	-	-	-	-
2	fraction	0,6±0,2	2,5±1,2	4±1,1	4±1,2	5±1,4	6±1,3
3	fraction	0,4±0,1	2,2±0,6	3.6±0,8	4±1,2	5±1,4	6±1,3
4	fraction	-	-	-	-	-	-
5	fraction	-	-	-	-	-	-

Table 6.

Effect of anticoagulant samples on activated partial thromboplastin time at different concentrations on platelet-poor mouse plasma in in vitro experiments.

thromboplastin time at different concentrations on platelet-poor mouse plasma in in vitro experiments

group	The drug	Anticoagulant concentration, units/time in sec					
		0,01	0,02	0,03	0,04	0,05	0,06
0	Geparin		54		180	>300	>300
1	Fraction	-	-	-	-	-	-
2	Fraction	-	60	-	200	320	350
3	Fraction	-	54	-	160	300	300
4	Fraction	-	-	-	-	-	-
5	Fraction	-	-	-	-	-	-

*control, with saline solution 20.8±2.0 sec.

As can be seen from the data presented in Table 7, thrombin time under the influence of both heparin and the studied samples of *A. persicus* changed much weaker than ACTH and therefore in the next series of experiments we used both heparin and the studied samples of preparations in doses an order of magnitude higher.

Effect of anticoagulant samples on thrombin time at different concentrations on platelet-poor mouse plasma in in vitro experiments

group	The drug	Anticoagulant concentration, units/time in sec				
		0,02	0,03	0,04	0,05	0,06
0	Geparin	19,7	25,7	133	>300	>300
1	Fraction	-	-	-	-	-
2	Fraction	21,7	27,7	153	>300	>300
3	Fraction	19,7	23,7	143	>300	>300
4	Fraction	-	-	-	-	-
5	Fraction	-	-	-	-	-

*control, with saline solution 15.6±1.0 sec.

As can be seen from the data presented in Table 8, there was a significant increase in APTB at concentrations of heparin 0.6-0.2 and *A. persicus* fraction. The greatest effect was observed at concentrations of 0.6 and 0.4 U (>300 sec).

Table 8.

Effect of anticoagulant samples on activated partial thromboplastin time at different concentrations on platelet-poor mouse plasma in in vitro experiments

group	The drug	Anticoagulant concentration, units/time in sec				
		0,2	0,3	0,4	0,5	0,6
0	Geparin	59,4	44,2	>300	126	>300
1	Fraction	-	-	-	-	-
2	Fraction	59,4	44,2	>300	128	>300

3	Fraction	49,4	40.4	>300	120	>300
4	Fraction	-	-	-	-	-
5	Fraction	-	-	-	-	-

*control, with saline solution 18.8±2.0 sec.

As can be seen from the thrombin time data in Table 9, heparin and salivary gland preparations of *A. persicus* ticks showed maximum effect at concentrations of 0.6-0.3ED.

Effect of anticoagulant samples on thrombin time at different concentrations on platelet-poor mouse plasma in in vitro experiments

group	The drug	Anticoagulant concentration, units/time in sec				
		0,2	0,3	0,4	0,5	0,6
0	heparin	147	>300	>300	>300	>300
1	fraction	-	-	-	-	-
2	fraction	147	>300	>300	>300	>300
3	fraction	147	>300	>300	>300	>300
4	fraction	-	-	-	-	-
5	fraction	-	-	-	-	-

*control, with saline solution 15.6±1.0 sec.

The presence of anticoagulant in the salivary gland of the studied bloodsucking tick *A. persicus was* shown by using modern tests. Moreover, this issue was studied not at the level of whole salivary gland extract of *A. persicus,* but after its preliminary fractionation into separate protein components. Based on these results it is possible to use the gland extract of ticks as a source of new anticoagulants, with their subsequent application in medical practice.

We also showed that not all fractions of proteins have anticoagulant properties, but two fractions (2 and 3 out of five). These results suggest that there are specific protein components in the mite gland extract that are responsible for their anticoagulant effect. Of course, it is of some interest to structurally and functionally characterize these protein components, which would allow their synthesis in order to use them in medical practice. In this aspect, this study is quite promising for the development of new generation anticoagulants.

CHAPTER VI. BIOACTIVE SUBSTANCES OF SALIVARY GLAND SECRETIONS OF IXODID TICKS AND TOXIC EFFECTS ON THE HOST ORGANISM

It is well known that ixodes ticks feed on the host's blood for several days, and their bite sites are accompanied by pathologic exudate, which is probably due to the impact of bioactive components of tick saliva. It is assumed that from the primary focus of the lesion toxic products of saliva get into the blood and, depending on the dosage, may cause symptoms of general poisoning of the organism (Pavlovsky and Alfeeva, 1941, 1949). In addition, in ixodid ticks, saliva has both anticoagulation and enzymatic properties (Kazakov et al., 2003; Motoyashiki et al., 2003; Tu et al., 2005).

Thus, there are direct and indirect indications in the available literature of the presence of toxins in the salivary glands of ticks. Nevertheless, to date, many species of ticks remain poorly studied in this regard. In particular, these include ixodes ticks inhabiting Uzbekistan.

4.1 Study of toxic effect of salivary gland secretions of *Hyalomma* and *Boophilus* mites on animal organism

The salivary gland (100 specimens each) of ticks was preliminarily separated from the body to obtain a homogenate. The obtained homogenate was left for 2 hours in a test tube for extraction of biologically active substances. Next, the homogenate was centrifuged for 10 min at 1000 rpm to obtain supernatant. The obtained supernatant was lyophilized according to the conventional method using liquid nitrogen. The obtained material was used for experimental studies. Extracts of three species of mites *Hyalomma asiaticum* *H. anatolicum* and *Boophilus calcaratus* were used.

Determination of toxicity on white mice and rats indicates almost the same sensitivity of these two species of warm-blooded animals to the poisonous secretion of ticks - the LD50 value for the saliva of the tick *Boophilus calcaratus* was 100 mg/kg, *Hyalomma anatolicum* - 170 mg/kg and *H. aciaticum* - 160 mg/kg. The venom of ixodid ticks has phospholipase, protease, hemorrhagic and neurotoxic activities, with the neurotoxic effect prevailing in the character of intoxication. The experimental animals die with the manifestation of paralysis, convulsion and respiratory distress. These data are a good enough reason to attribute the studied ixodes ticks to the group of poisonous arthropods (Kazakov, 2010).

Venom of the mite *Boophilus calcaratus* at relatively low concentrations (up to 170 µg/mL protein) caused stimulation of Mx respiration in metabolic states 2 and 4, but had almost no effect on oxygen consumption in state 3, which was accompanied by a decrease in the value of DA and ADP/O. However, a venom dose of 250 µg/mL almost completely dissociated the oxidative phosphorylation of Mx. As follows from the data obtained, EDTA at a concentration of 5 mM only partially relieved the effect of whole venom. Thus, if the venom at a dose of 250 µg/mg protein inhibited mitochondrial respiration in state 3 by 60% and the value of DC by 62%, in the presence of EDTA this inhibition was 30 and 45%, respectively, of the control level. Consequently, in addition to phospholipase A, other membrane-active components capable of influencing respiration and oxidative phosphorylation parameters are present in the composition of mite venom.

When fractionated on Sephadex G-75, the venom of the tick *Boophilus calcaratus* was divided into 4 protein fractions. I fraction of venom at a dose of 500 µg/mL causes inhibition of Mx respiration in state 3 and to a small extent - increase of respiration in state 4.

In the case of II fraction of tick salivary gland secretion (250 µg/mL), a sharp increase in the respiration rate of Mx in state 4 and a dissociative effect on oxidative phosphorylation were observed. This fraction simultaneously increased Mx respiration in state 3 as well to a small extent (up to 25%). In these experiments, the increase in respiration in state 4 caused by the II fraction of tick venom deserves special attention. According to preliminary data, this fraction has phospholipase activity, which may cause such an effect.

In experiments on Mx using the III fraction of tick venom, we noted its low efficiency, which is expressed only in a slight inhibition of Mx respiration at a concentration of 500 µg/mL of this fraction. IV fraction of poisonous secretion of salivary glands of ticks is of special interest. At its action stimulation of Mx respiration is noted, and a noticeable effect can be detected already at a concentration of 50 µg/mL, and the magnitude of the effect continues to grow as the concentration increases up to 500 µg/mL.

In general, the extract of salivary glands of ticks and fractions isolated from it are very similar to caracourt venom by the nature of the action at the Mx level (Kazakov et al., 2003; Kazakov, 2010).

As follows from the above, it is of some interest to compare the membrane-active properties of the venom secretion of ixodal ticks.

In experiments on BLM it was shown that the addition of 10 µg/mL of *Boophilus calcaratus* mite venom to one of the compartments of the experimental cell led to an increase in the conductivity of BLM by several orders of magnitude. Further increase in the amount of venom was accompanied by a decrease in the average membrane lifetime, i.e., the membrane destabilized and destroyed. One of the reasons for such labilization and its destruction may be related to the presence of phospholipase activity in the studied venom.

The study of the conductivity of modified membranes in the mode of membrane potential fixation showed that the action of venom in small amounts (2-M µg/mL) causes a jump-like increase in BLM conductivity (Fig. 6). Consequently, the mite venom really forms discrete ion-conducting channels on BLM.

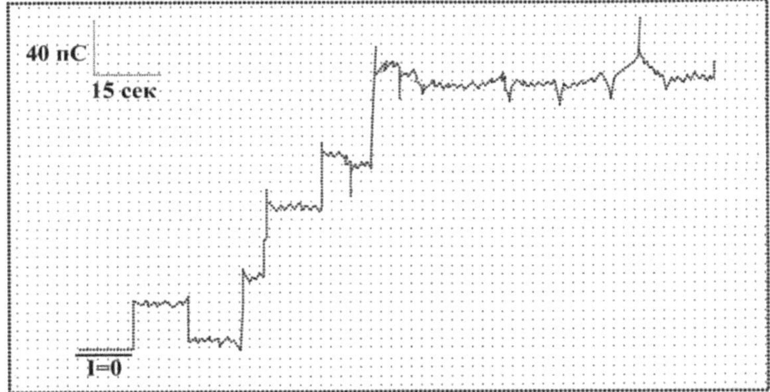

Figure 6. Current recordings through channels formed by *Boophilus calcaratus* tick venom *(2+4* μg/mL). Medium: 100 mm sas, 5 mm Tris-NS1, ph 7.5.

In order to identify the channelformer component, the venom of the tick *B.calcaratus* was fractionated on Sephadex G-75. Under these conditions, the whole venom was divided into 4 fractions containing protein components of the following molecular masses: fraction I - 100 kDa and below; fraction II - 60-70; fraction III - 30-40; fraction IV - 5-6 kDa. Among these fractions, only fraction IV had pronounced membrane-active properties, causing a discrete increase in BLM conductance, i.e. formed single channels.

Thus, channels induced by the IV fraction of tick venom on BLM have cation selectivity, matching in all characteristics with ion channels formed by whole venom as well.

It should be emphasized that both phospholipase and neurotoxin of tick venom may differ in structure from more active analogs of venom from other animals, which is undoubtedly an important molecular toxicological aspect requiring further special studies. The anticoagulant was purified by a two-step method and Sephadex G-75 column chromatography was used. The presence of anticoagulant in the venom fractions was verified with measurement of plasma clotting time. The active fractions were further purified by DEAE - SEPHADEX A-25 column chromatography. The crude venom was purified until obtained in homogeneous form against weight markers. The molecular weight of the kalkaratin component was 14,500 Da. This anticoagulant can be used in practical medicine for the prevention of cardiovascular diseases.

42

CONCLUSION

Ticks of the superfamily Ixodoidea are a group of highly specialized blood-sucking arthropods. Ixodoid ticks are distinguished by their gigantic body size, absorption of huge volumes of blood and lysed tissues of vertebrates, and exceptional fecundity. The perfection of adaptations of argas and ixoid mites to temporary parasitism on terrestrial vertebrates clearly distinguishes them from other groups of arthropods.

The practical role of ixodoid ticks is determined by their participation in the transmission of pathogens of the most dangerous vector-borne infections and infestations of humans and farm animals, as well as by the severe consequences of their parasitization itself. Improvement of methods and means of protection against attacks of bloodsucking ticks, their control and pathogens transmitted by them stimulate research on ecology, faunistics and taxonomy of ectoparasites, as well as on various aspects of their veterinary and medical importance. Along with these traditional directions, works on morphology, biology, and life cycles of argasids are gaining great weight. Experimental study of various aspects of mite life activity requires comprehensive studies carried out at the modern level on information about the functioning of the parasitic system "mites - vertebrates" in specific areas.

The choice of the object of research is dictated by the above circumstances.

In argasid ticks, most species of which are characterized by the nesting-nesting type of parasitism, the amount of blood absorbed once and fecundity significantly exceed the similar indicators among gamazid-hematophagous mites, but are inferior to those of ixodid ticks with a more perfect pasture-understory type of parasitism (Balashov, 1967).

The studies have shown a wide distribution of Argas mites in terrestrial cenoses of Uzbekistan as ectoparasites of domestic and wild animals. They are represented by 6 species belonging to two genera and two subfamilies. It was found that representatives of the genus *Argas* are dominant parasites mainly of domestic chickens, turkeys and pigeons. The total infestation of birds ranges from 5.3 - 84.6 % with the intensity of infestation 9 - 54 exv.

In most cases, mites of the genus *Argas do* not show strict specificity to certain species or groups of animal hosts. They are found both in birds and mammals. A similar situation was observed with populations of species of the genus *Ornithodoros*, which can parasitize different groups of animals. In this respect, the specificity of *Ornithodoros lahorensis* mites to parasitize only sheep should be noted.

The complex of argas mites as ectoparasites has a serious impact on the host organism. The high feeding intensity of mites leads to significant blood loss by vertebrates. Bioactive components of salivary gland secretions injected by ticks into the host organism may be the cause of poisoning of animals with subsequent consequences. Moreover, the incidence of argas mites attacks on humans is increasing, which emphasizes their epidemiological role in natural and urbanized areas.

An extremely important aspect of the work is the study of bioactive substances of salivary gland secretions of Argas mites. As experimental studies have shown, phospholipase A, protease and other active substances have been isolated in the composition of salivary gland secretions of species of the genus *Argas*. Anticoagulant and relaxant properties of some fractions isolated from the saliva of mites were determined.

The results of the study of bio-damaging bioactive substances of *Boophilus calcaratus* salivary gland secretions also demonstrate certain effects on host mitochondrial membranes. The venom of *B. calcaratus* ticks at low concentrations (up to 170 µg/mL protein) causes stimulation of mitochondrial respiration in states 2 and 4, but does not affect oxygen consumption in state 3, which is accompanied by a decrease in DA and ADP/O values. However, a venom dose of 250 µg/mL completely suppresses ATP-synthesizing functions of mitochondria, and this effect of venom is prevented when EDTA is added to the medium. This indicates a significant role of phospholipases and other hydrolases and probably presynaptic neurotoxin included in the studied venom and their dependence on the level of ionized Ca^{2+} in the medium.

It was found that the venom of *B. calcaratus* ticks has the ability to induce single ion channels on BLM and effectively modify membrane-dependent mitochondrial functions (Kazakov, 2010).

During fractionation on Sephadex G-75 the venom of *B.calcaratus* ticks is divided into 4 protein fractions. At the same time, only the low-molecular-weight IV fraction forms channels on BLM. This low-molecular-weight peptide is probably the main channel-forming component of the whole venom.

By fractionation by two-step column chromatography on Sephadex G-75, and DEAE-Sephadex A-25, a purified component, calcaratine, with a molecular mass of 14,500 Da, was isolated from the venom of *B.* calcaratus ticks. (Tu et al., 2003; Kazakov, 2010).

The importance of ticks as obligate hematophages and carriers of pathogenic agents of wild and domestic animals, as well as humans is very high. This necessitated the development of effective methods and means of tick control. In this respect, the results of the conducted experimental and production tests of new acaricides deserve special deduction. The tested preparations, in particular, acaricides of plant and fungal origin open wide prospects for the creation of new classes of drugs and their wide use in the fight against bloodsucking mites - ectoparasites of animals and humans.

Acknowledgement

The authors are grateful to the staff of the Laboratory of General Parasitology of the Institute of Botany and Zoology of the Academy of Sciences of the Republic of Uzbekistan and the Laboratory of Chemistry of Proteins and Peptides of the Institute of Bioorganic Chemistry of the Academy of Sciences of the Republic of Uzbekistan for their assistance in conducting experimental studies. The work was carried out within the framework of the fundamental projects of the Academy of Sciences of the Republic of Uzbekistan FA-F5-T230 and EF5-FA-O17793 (2012-2016).

LITERATURE

1. Abdurasulov Sh.A. Development of culture strain *Theileria annulata* TAU-219 in ticks of the genus *Hyalomma*: dissertation ... ... Candidate of Biological Sciences. - Tashkent, 2006. - 23 c.

2. Agrinsky N.I. Insects and mites harming agricultural animals. - Moscow, 1962. - 288 c.

3. Alekseev A.N., Kondrashova Z.N. The organism of arthropods as a habitat for pathogens. - Sverdlovsk: "Nauka", 1985. - 177 c.

4. Alexeev A.N. The system of tick pathogen and its emergent properties. - St. Petersburg, 1993. - 205 c.

5. AmosovaL .I. Comparative electron-microscopic study of salivary glands of parasitic and free-living parasitiform mites // Proceedings of the IV All-Russian Congress "Parasitology in the XXI century - problems, methods, solutions". - St. Petersburg, 2008. - Vol. 1. - C. 15-19.

6. Balashov Y.S. Blood-sucking ticks (Ichosiisiae) - carriers of human and animal diseases. - Leningrad: "Nauka", 1967. - 320 c.

7. Balashov Y.S. Parasite-host relations of arthropods with terrestrial vertebrates. - Leningrad, Izd. "Nauka", 1982. - C. 345.

8. Baluda V.P. - Laboratory methods of investigation of the hemostasis system. Tomsk, 1980, 310 p.

9. Belozerov V.N. Dynamics of gas exchange during the development of ixodial ticks (Acarina, Ixodidae) // Journal Entomological Review. - Moscow, 1966. - № 3. - C. 509-520.

10. Bogdanova E.N.. Processes of synanthropization of ticks and their epidemiological significance // Proceedings of the IV All-Russian Congress "Parasitology in the XXI century problems, methods, solutions" - St. Petersburg, 2008. - Vol. 1. - C. 80-84.

11. Galuzo I.G. Argas mites (Argasidae) and their epizootologic significance. Alma-Ata, 1957.-129 p.

12.   Ginetsinskaya T.A., Dobrovolsky A.A. Private Parasitology. Parasitic worms, mollusks and arthropods. - Moscow, "Vysshaya Shkola", - 1978. Kn. 2. - P. 168-279.

13.   Kazakov I., Abubakirova M., Reimbayeva R.,K.T., Almatov., Azimov D.A., Anthony T. Tu. Effect of different fractions of venom of ticks *Boophilus calcaratus* on oxidative phosphorylation of rat liver mitochondria // Izvestiya Vuzov. - Tashkent, 2003. - №1 - 2. - C. 94 - 98.

14.   Kazakov I., Abubakirova M., Reimbayeva R., Azimov D.A., Mirkhodjaev U.Z. Poisonous animals of Uzbekistan and prevention of their bites // Nukus, 2010. C. - 56.

15.   Kazakov I. Comparative characterization of membrane activity of secretions of poisonous animals of Uzbekistan: autorref. dissertation of doctor of biological sciences. -
Tashkent, 2010. - 50 c.

16.   Kennedy K. Ecological parasitology. - Moscow, 1978. - 232 c.

17.   Kogan A.V., Shitov S.I. Technique of physiological experiment. - Moscow, "Vysshaya Shkola", - 1967. -565 c.

18.   Kuzibayeva H. Burrowing mites of the genus *Alectorobis* and their importance as vectors of spirochetes in Uzbekistan: autoref.   diss....
Candidate of Biological Sciences. -
Tashkent, 1964. -21 c.

19.   Kuzibaeva H. On distribution and ecology of the mite of the genus *Argas* Latr., 1796 in the Fergana Valley // In the collection Ecology and biology of animals of Uzbekistan. Tashkent, 1969.-C. 75-478.

20.   Kuklina T.E. Fauna of ixodial ticks of Uzbekistan. - Tashkent, "Fan", - 1976. - 146 c.

21.   Magomedova S.A. Ticks and insects - carriers of pathogens of human infectious diseases in lowland and foothill belts of Dagestan: autoref.dissertation of Candidate of Biological Sciences. - Makhachkala, 2004. -25 c.

22. Malygin V.A., Alekseev A.N. Changes in gas exchange in mites *Hyalomma asiaticum P.Sch. et Schl.* 1929 depending on environmental conditions // Zoological Journal. - Moscow, 1987. №39(2). - C. 297-299.

23. Mirzaeva A.U., Kazakov I., Abubakirova M.I., Bekmirzaev M.H., Bekmuradova N.T. Gas exchange in ticks of the family Argasidae // UzMU habarlari. - Toshkent, 2011. - C. 130-131.

24. Muratbekov Y.M. To the question of zoogeography of ixodid ticks of Tashkent region // Proc. of the Institute of Zoology and Parasitology. Institute of Zoology and Parasitology. Sb. of parasitology. - Tashkent, 1954. - C. 65.

25. Muratbekov Y.M. To the question about the fauna of external parasites of domestic and wild animals in the Amu Darya delta village // Materials of the Conf. on productive forces of Uzbekistan. -Tashkent, 1959. 10 - C. 25-28.

26. Muratbekov Y.M. Fauna of ectoparasites in the foothills of Khavast district // Proc. of the Uzbek Academy of Sciences. Uzbek. Institute of Epidemiology and Microbiology named after A.M. Mukhtarov, - 1949. 10th Anniversary of Soviet Medicine, - 1949. - C. 33-34.

27. Muhammedkulov M. To the fauna of ixodid and argas ticks of birds of the Zarafshan valley // Second Acarological Meeting. Kiev: Naukova Dumka, 1970.-C. 90-92.

28. Nadyrov S.A. Dynamics of attack of blood-sucking ticks on cattle in conditions of Tashkent region // In Collection: Vopros. biol. and regional medicine. - Tashkent: Academy of Sciences of the UzSSR, 1962. - Issue. 3. - C. 37.

29. Orlov B.N., Gelashvili D.B. Zootoxicology: poisonous animals and their poisons. - M.: Vysshaya Shkola, 1985. - C. 75-92.

30. Pavlovsky E.N., Alfeeva SP. Pathologic and histologic changes in the skin of cattle at the bite of the tick Ixodes ricinus // Proc. of the Military Medical Academy. Military Medical Academy. acad. named after S.M.Kirov.- 1941.- [1] 25.- P. 153-160.

31. Pavlovsky E.N., Alfeeva SP. Comparative pathology of mammalian

skin in case of tick bite. The effect of the bite of Nuaiotta mites on the skin of the bull, cow, goat and dog // Izv. AS USSR, ser.biol.- 1949.- № 6.-P.709-715.

32. Podboronov V.M., Podboronov A.M. Lysozyme and other antibacterial factors of parasitic arthropods and their effect on pathogenic microorganisms. - Moscow, 1993. - C. 291.

33. Pospelova - Shtrom M.V. Ticks - ornithadorines and their epidemiologic significance. - Moscow, Izd, Academy of Sciences of the USSR, 1953. - 102 c.

34. Rasulov I.H., Askarkhodjaeva Abdurasulov S.A. Experiments on the invasion of the tick *Hyolamma anatolicum*, attested strain *Theileria annulata* TAU-219 // "Problems of veterinary arachnology in the new millennium" Proc. of the Interdisciplinary Scientific Conference - Tyumen, 2001. - C. 218-221.

35. Rasulov I.H., Abdurasulov S.A., Nazrullaeva M.F. Ixodofauna and taxonomy of ticks vectors of bovine pyroplasmidosis in irrigated areas of Syrdarya and Jizzak regions // Actual problems of parasitology. Materials of Resub. scientific-practical conf. - Karshi, 2003. - C.110 - 114.

36. Serzhanov O.S. Ixodes ticks of Karakalpakstan and their epidemiologic and epizootologic importance: autoref. dissertation ...kand.biol.nauk. - Nukus, 1964. -23 c.

37. Stefanova A.V., Preclinical study of harmlessness of drugs. Preclinical study of antianemic agents, anticoagulants and fibrinolytics, State Pharmacological Center, ed. - Kiev , 2002, P. 310-326.

38. Uzakov U.Y. Ixodes mites of Uzbekistan. Tashkent: Fan, 1972. - 302 c.

39. Uzokov U.Y. Ixodes mites of Uzbekistan. Tashkent: Fan, 1974. - 304c.

40. Shashina N.I. Scientific bases of development of means of individual protection of people from attack of ixodovyh ticks - carriers of dangerous diseases: abstract of dissertation .cand.biol.nauk. - Moscow, 2007. -4-11 c.

41. Shevkoplyas V.N. Thermoaerosols "Tatsik" in protection of farm animals from temporary parasites: mosquitoes, flies, lice, chicken and Persian mites: abstract of dissertation. Candidate of Biological Sciences. - Moscow, 2002. - 5-9 c.

42. Abdel-Shafy, S., Soliman M.M., Habeeb S.M. *In vitro* acaricidal effect of some crude extracts and essential oils of wild plants against certain tick species // J. Acarol. Acarol. XLVII. - Iran, 2007. - V. (1-2). - P. 33-42.

43. Abduz Zahir, A., Abdul Rahuman A., Kamaraj C., Bagavan A., Elango G., Sangaran A., Senthil Kumar B.. Laboratory determination of efficacy of indigenous plant extracts for parasites control // Parasitol. Res. - 2009. - V. 105. - P. 453-461.

44. Ahmed J., Alp H., Aksin M., Seitzer U. Current status of ticks in Asia // Parasitol. Res. - 2007. - V. 101(2). - P. 159-162.

45. Al-Rajhy D.H., Al-Ahmed A.M., Hussein H.I., Kheir S.M., Acaricidal effects of cardiac glycosides, azadirachtin and neem oil against the Camel tick, *Hyalomma dromedarii* (Acari: Ixodidae) // Pest. Manag. Sci. - 2003. - V. 59. - P. 1250-1254.

46. Anderson J.F., Magnarelli L.A. Biology of Ticks // Infect. Disease Clinics of North America. - 2009. - V. 22(2). - P. 195-215.

47. Andrade B.B., Teixeira C.R., Barral A., Barral-Netto M. Haematophagous arthropod saliva and host defense system: a tale of tear and blood //Anais da Academia Brasileira de Ciencias. - 2005. - V. 77(4). - P. 665-693.

48. Andrew Y.L., Davey R.B., Miller R.J., George J.E.. Detection of characterization of amitraz resistance in the southern cattle tick, *Boophilus microplus* (Acari, Ixodidae) // J. Med. Med. Entomol. - 2004. - V. 41. - P. 193-200.

49. Baker A.S., Craven J.C. Hecklist of the mites (Arachnida: Acari) associated with bats (Mammalia: *Chiroptera*) in the British Isles // Syst. Appl. Acarol. Spec. Publ. - 2003. - V. 14. - P. 1-20.

50. Bergstrom, S., Haemig, P.D., Olsen B. Increased mortality of black-browed albatross chicks at a colony heavily-infested with the tick *Ixodes* uriae // Int. J. Parasitol. - 1999. - V. 29. - P. 1359-1361.

51. Bowman A.S., Sauer J.R.. Tick salivary glands: function, physiology, and future // Parasitol. - 2004. - V. 129. - P. S67-S81.

52. Bowman A.S. Tick salivary prostaglandins: Presence, origin and significance // Parasitol., 1996. - V. - 12. - P. 388 - 396.

53. Brossard M., Wikel S.K.. Tick immunobiology // Parasitol. - 2004. - V. 129. - P. S161-S176.

54. Cairoll J.F., Solberg V.B., Klun J.A., Kramer M., Debboun M., Comparative activity of Deet and AI3-37220 Repellents against the ticks *Ixodes ascapularis* and

Amblyomma americanum (Acari: Ixodidae) in Laboratory bioassays // J. Med. Med. Ent. - 2004. - V. 41. - P. 249-254.

55. Chandler D., Davidson, G., Pell J.K., Ball B.V., Shaw K., Sunderland K.D. Fungal biocontrol of Acari // Biocontrol science and technology. - 2000. - V. 10. - P. 357-384.

56. Chinikara S., Ghiasia S.M., Hewsonc R., Moradia M., Haeri A. Crimean- Congo hemorrhagic fever in Iran and neighboring countries // J. Clinical Virol. Clinical Virol. - 2010. - V. 47. - P. 110-114.

57. Chmelar J., Oliveira C.J., Rezacova P., Francischetti I.M., Kovarova Z., Pejler G., Kopacek P., Ribeiro J.C., Mares M., Kopecky J., Kotsyfakis M. A tick salivary protein targets cathepsin G and chymase and inhibits host inflammation and platelet aggregation // Blood. - 2011. - V. 117(2). - P. 736-744.

58. Choudhary R.K., Asanthi V.C., Latha B.R., John L. *In vitro* effect of Nicotiana tabacum aqueous extract on Rhipicephalus haemaphysaloides ticks // Ind. J. Anim. Sci. - 2004. - V. 74(7). - P. 730-731.

59. Chungsamarnyart N., Jansawan W., Effect of *Tamarindus indicus L.* against the *Boophilus microplus* // Kasetsart J. Narur. Sci. Kasetsart Umv. -

2001. (Bangkok, Thailand). - V. 35(1). - P. 34-39.

60. Chungsamarnyart N., Ratanakreetakul C., Jansawan W. Acaricidal activity of the combination of plant crude-extracts to tropical cattle ticks // Kasetsart J. Narur. Sci. - 1994. - V. 28(4). - P. 649-660.

61. Corral-Rodriguez M.A., Macedo-Ribeiro S., Pereira B.P.J., Fuentes-Prior P. Tick-derived Kunitz-type inhibitors as antihemostatic factors // Insect. Biochem. Mol. Biol. - 2009. - V. 39(9). - P. 579-95.

62. Dautel H., Test systems for tick repellents // J. Med. Med. Mic. - 2004. - V. 293. - P. 182-188.

63. Dennis D.T., Piesman J.F., Overview of tick-borne infections of humans, In tick-borne diseases of humans, D.T.D. Goodman J.L., Sonenshine D.E.. Ed. Am. Soc. Micr. Press: Washingtion, DC. - 2005. - P. 3-11.

64. Donat J.R., Donat J.F.. Tick paralysis with persistent weakness and electromyographic abnormalities // Arch. Neurol. - 1981. - V. 38. - P. 59-61.

65. Drummond R.O. Tick-borne livestock diseases and their vectors. Chemical control of ticks // Wld. Anim. Rev. (FAO). - 1983. - V. 36. - P. 28-33.

66. Durden L.A., Mclean R.G., Oliver J.H., Ubico S.R., James A.M., Ticks, Lyme disease spirochaetes, trypanosomes and antibody to encephalitis viruses in wild birds from coastal Georgia and South Carolina // J. Parasitol. Parasitol. - 2001. - V. 83. - P. 11781182.

67. Dusba'bek F., Rupes V., Sbreveimek P., Zahradni'ckova' H. Enhancement of permethrin efficacy in acaricide - attractant mixtures for control of the fowl tick *A. persicus* (Acari: Argasidae) // Exp. Appl. Acarol. - 1997. - V. 21(5). - P. 293305(13).

68. Dutra V., Nakazato L., Broetto L., Schrank S.I., Vainstein H.M., Schrank A. Application of representational difference analysis to identify sequence tags expressed by *Metarhizium anisopliae* during the infection process of the tick *Boophilus microplus* cuticle // Res. Microbiol.- 2004. - V. 155. - P. 245-251.

69. Estrada-Pena A., Bouattour A., Camicas J.L., Walker A.R.. Ticks of domestic animals in the Mediterranean region. A guide to identification of species // Univ. Zaragoza, Spain. - 2004. - P. 131.

70. Estrada-Pena A., Mangold A.J., Nava S., Venzal J.M. A review of the systematics of the tick family Argasidae (*Ixodida*) // Acarol. - 2010. - V. 50(3). - P. 317-333.

71. Estrada-Pena A., Venzal J.M., Gonzalez-Acuna D., Mangold A.J., Guglielmone A.A. Notes on new world persicargas ticks (Acari: Argasidae) with description of female *Argas* (*P.*) *keiransi* // J. Med. Med. Entomol. - 2006. - V. 43(5). - P. 801-809.

72. Fernandes, F.F., Alessandro W.B.D., Edmeia F.P.S. Toxicity of extract of *Magonia pubescens* (Sapindales: *Sapindaceae*) St.Hil. to control the brown Dog Tick, Rhipicephalus sanguineus (Latreille) (Acari: *Ixodidae*) Neotro // Entom. - 2008. - V. 37(2). - P. 205-208.

73. Foil L.D., Coleman P., Eisler M., Fragoso-Sanchez H., Garcia-Vazques Z., Guerrero F.D., Jonsson N.N., Langstaff I.G., Li A.Y., Machila N. Langstaff I.G., Li A.Y., Machila N. Factors that influence the prevalence of acaricide resistance and tick-borne diseases // Veter. Parasitol. - 2004. - V. 125. - P. 163-181.

74. Francischetti I.M. Sa-Nunes A., Mans B.J., Santos I.M., Ribeiro J.M. The role of saliva in tick feeding // Front. Biosci. - 2009. - V. 14. - P. 2051-2088.

75. Frazzon G.A., Vaz Jr. S.I., Masuda A., Schrank A., Vainstein H.M. *In vitro* evaluation of *Metarhizium anisopliae* isolates to control the cattle tick *Boophilus microplus* // Vet. Parasitol. - 2000. - V. 94. - P. 117-125.

76. Frenandez-Ruvalcaba M., Cruz-Vazquez C., Solano-Vergara J., Garcia-Vazquez Z. Anti-tick effects of Stylosanthes humilis and Stylosanthes hamata on plots experimentally infested with *Boophilus microplus* larvae in Morelos, Mexico // Hxp. Appl. Acarol. - 1999. - V. 23(2). - P. 171-175.

77. Fry B.G., Roelants K., Champagne D.E., Scheib H., Tyndall J.D., King

G.F., Nevalainen T.J., Norman J.A., Lewis R.J., Norton R.S., Renjifo C., Rodriguez de la Vega R.C. The Toxicogenomic multiverse: Convergent recruitment of proteins into animal venoms // An. Rev. Genom. Human Gen. - 2009. - V. 10. - P. 483-511.

78. Fukumoto S., Sakaguchi T., You M., Xuan X., Fujisaki K. Tick troponin I-like molecule is a potent inhibitor for angiogenesis // Microvasc. Res. - 2006. - V. 71. - P. 218-221.

79. Ghadeer M., Blank G., Han J.H., Hydamaka A., Holley R.A. Effectiveness of trisodium phosphate, lactic acid and commercial antimicrobials against pathogenic bacteria on chicken skin // Food Protection Trends. - 2005. - V. 25. - P. 351-362.

80. Gindin G., Samish M., Zangi G., Mishoutchenko A., Glazer I. The susceptibility of different species and stages of ticks to entomopathogenic fungi // Exp. Appl. Acarol. - 2001. - V. 8. - P. 283-288.

81. Gothe R. Tick paralyses: Reasons for appearing during *Ixodid* and *Argasid* feeding. In Current topics in vector research (ed. Harris K.F.) // Praeger Publishers (New York). - 1984. - V. 2. - P. 199-223.

82. Gothe R. Zecken toxikosen hieronyms // Munich, JSBN 3-88751-083-17. - 1999. - P. 377.

83. Gothe R., Neitz A.W.. "Tick paralysis: pathogeneses and etiology," in Advances in disease vector research, Harris K.F. Ed. // Springer (New York, NY, USA). - 1991. - V. 8. - P. 177-204.

84. Guglielmone A.A., Robbins R.G., Apanaskevich D.A., Petney T.N., Estrada- Pena A., Horak I.G., Shao R., Barker S.C. The Argasidae, Ixodidae and *Nuttalliellidae* (Acari: Ixodidae) of the world: a list of valid species names // Zootaxa. - 2010. - V. 2528. - P. 1-28.

85. Habbib S.M., Sewify G.H.. Biological control of the fowl tick *Argas (Persicargas) persicus* (Laterreille) by the entomopathogenic fungi *Beauveria bassiana* and *Metarhizium anisopliae* // Egypt. J. Biol. Pest. Contr. - 2002. - V. 12. - P. 1.

86. Habbib S.M. Ethno - Veterinary and medical knowledge of crude plant Extracts and its Methods of Application (Traditional and Modern) for Tick Control // World Appl. Scien. Scien. J. - 2010. - V. 11(9). - P. 1047-1054.

87. Harrow I.D., Gration K.A. F., Evans N.A. Neurobiology of arthropod parasites // J. Parasitology, 1991. Parasitology, 1991. V. 102. - P. 59-69.

88. Hassanain M.A., Garhy M.F., Abdel-Ghaffar F.A., El-Sharaby A., Abdel Megeed K.N.. Biological control studies of soft and hard ticks in Egypt. I The effect of *Bacillus thuringiensis* varieties on soft and hard ticks (Ixodidae) // Parasitol. Res. - 1997. - V. 83(3). - P. 209-213.

89. Hepburn N.J., Williams A.S., Nunn M.A., Chamberlain-Banoub J.C., Hamer J. et al. *In vivo* characterization and therapeutic efficacy of a C5-specific inhibitor from the soft tick *Ornithodoros moubata* // J. Biol. Biol. Chem. - 2007. - V. 282. - P. 82928299.

90. Hillyard P.D. Ticks of North-West Europe. Synopses of the British Fauna (New series) No. 52 (ed. by R.S.K. Barnes and J.H. Crothers) Field Studies Council, Shrewsbury. - 1996. - P. 178.

91. Hornbostel V.L., Ostfeld R.S., Benjamin M.A. Effectiveness of *Metarhizium anisopliae* (*Deuteromycetes*) against *Ixodes scapularis* (Acari: *Ixodidae*) engorging on *Peromnyscus leucopus* // J. Vect. Vect. Ecol. - 2005. - V. 30. - P. 91-101.

92. Jansawan W., Jittapalapong S., Jantaraj N. Effect of Stemona collinsae extract against cattle ticks (*Boophilus microplus*) Kasetsart // J. Narur. Sci. - 1993. - V. 27(3). - P. 336-340.

93. Jongejan F., Uilenberg G. The global importance of ticks // Parasitol. - 2004. - V. 129. - P. S3-S14.

94. Kaaya G.P. Prospects for innovative methods of tick control in Africa // Ins. Scien. Appl. - 2003. - V. 23. - P. 59-67.

95. Kaaya G.P. The potential for anti- tick plants as components of an integrated tick control strategy // Ann. New York Acad. Sci. (USA). - 2000. - V. 916. - P. 576582.

96. Kaaya G.P., Mwangi EN., Malonza M.M. Acaricidal activity of Margaritaria discoidea (Euphorbiaceae) plant extracts against the ticks *Rhipicephalus appendiculatus* and *Amblyomma variegatum* (Ixodidae) // Intern. J. Acarol., - 1995. - V. 21(2). - P. 123-129.

97. Kandil O.M., Habeeb S.M., Nasser M.M. Adverse effect of Sorghum bicolor, Sea anemone, Cynobateria spp and Simmondsia chiinensis (Hohoba) extracts on reproductive physiology of the adult female tick, *Boophilus annulatus* // Assiut. Vet. Med. J. - 1999. - V. 42. - P. 29-37.

98. Kanturk Yigit G. An example of tick-Crimean Congo hemorrhagic fever (CCHF) in Eflani district, Karabuk, Turkey // Scientific Research and Essays. - 2011. - V. 6(11). - P. 2395-2402.

99. Kazakov I., Abubakirova M.E.. On some biophysical traits of Ixodidae mites venoms // Abstracts 27th Pakistan congress of Zoology (International Congress) February 27 to March 1, 2007. Bahauddin Zakariya university, Multan. - Hamdard Pakistan, 2007. P. 1.

100. Kazimirova M. Pharmacologically active compounds in tick salivary glands // Makarov S.E., Dimitrijevic R.N. (Eds.) Inst. Zool., Belgrade; BAS, Sofia; Fac. Life Sci., Vienna; SASA, Belgrade & UNESCO MAB Serbia. Vienna -Belgrade. - Sofia, 2008. Monographs. 12. - P. 281-296.

101. Khater H.F., Ramadan M.Y.. The acaricidal effect of peracetic acid against *Boophilus annulatus* and *A. persicus* // Acta Scientiae Veter. - 2007. - V. 35(1). - P. 29-40.

102. Khudrathulla M.D., Jagannath M.S. Biocontrol of ixodid ticks by forage legume Stylosanthes scabra (Vogel) // Indian J. Animal. Sci. - 1998. - V. 68(5). - P. 428-430.

103. Kinsey A.A., Durden L.A., Oliver Jr. Tick infestation of birds in coastal Georgia and Albama // J. Parasitol. Parasitol. - 2000. - V. 86. - P. 251-254.

104. Knipling E.F., Steelman C.D. Feasibility of controlling *Ixodes scapularis* ticks (Acari: Ixodidae), the vector of Lyme Disease, by parasitoid augmentation // J. Med. Med. Entomol. - 2000. - V. 37. - P. 645-652.

105. Koh C.Y., Kazimirova M., Trimnell A., Takac P., Labuda M. et al. Variegin, a novel fast and tight binding thrombin inhibitor from the tropical bont tick // J. Biol. Biol. Chem. - 2007. - V. 282. - P. 29101-29113.

106. Kumar R., Chauhan P.P., Agrawal R.D., Shankar D. Efficacy of herbal ectoparasiticidal compound AV/EPP/14 against lice and tick infestation on buffalo and cattle // J. Vet. Vet. Parasit. - 2000. - V. 14(1). - P. 67-69.

107. Lagos J.C., Thies R.E.. Tick paralysis without muscle weakness // Arch. Neurol. - 1969. - V. 21. - P. 471-474.

108. Lundh, J., Wiktehus D., Chirico J. Azadirachtin-impregnated traps for the control of Dermanyssus gallinae // Vet. Parasitol. - 2005. - V. 130. - P. 337-42.

109. Lysyk T.J., Majak W., Veira D.M. Prefeeding Dermacentor andersoni (Acari: Ixodidae) on cattle with prior tick exposure may inhibit detection of tick paralysis by using hamster bioassay // J. Med. Med. Entomol. - 2005. - V. 42(3). - P. 376-382.

110. Magano S.R., Thembo K.M., Ndlovu S.M., Makhubela N.F.. The anti-tick properties of the root extracts of Senna italica subsp. Arachoides // Afnc. J. Biotech. - 2008. - V. 7(4). - P. 476-481.

111. Mans B.J., Gothe R., Neitz A.W.. Biochemical perspectives on paralysis and other forms of toxicoses caused by ticks // Parasitol. - 2004. - V. 129. - P. S95-S111.

112. Mans B.J., Gothe R., Neitz A.W.H.. Biochemical perspectives on paralysis and other forms of toxicoses caused by ticks // Parasitol. - 2002. - V. 129. - P. S95- S111.

113. Maritz C., Louw A. I., Gothe R., Neitz A.W.. Neuropathogenic properties of *Argas* (*Persicargas*) *walkerae* larval homogenates // Comp. Biochem. Physiol. - 2001. - V. A 128. - P. 233-239.

114. Maritz-Olivier C., Stutzer C., Jongejan F., Neitz A.W., Gaspar A.R.. Tick antihemostatics: targets for future vaccines and therapeutics // Trends Parasitol. - 2007. - V. 23. - P. 397-407.

115. Martin C.J., Torres F., Quinones W., Cheverri F. Phytochem is try and evaluation of biological action of Polygonum punctatum // Rev. Eatinoamericana de Quimica. - 2001. - V. 29(2). - P. 100-107.

116. Masina S., Broady K.W.. Tick paralysis: development of a vaccine // Inter. J. Parasitol. - 1999. - V. 29. - P. 535-541.

117. Massoud A.M., Kutkat M.A., Abdel-Shafy S., El-Khateeb R. Acaricidal efficacy of Myrrh commiphora molmol on the fowl tick *A. persicus* (Acari: Argasidae) // J. Geophys. Egypt. Soci. Parasitol. - 2005. - V. 35(2). - P. 667-686.

118. Matovu H., Olila D., Acaricidal activity of Tephrosia vogelii Extracts on Nymph and Adult Ticks // Int. J. Trop. Med. - 2007. - V. 2(3). - P. 83-88.

119. Mawela K.G. The toxicity and repellent properties of plant extracts used in ethnoveterinary medicine to control ticks // Magister Sc. Thesis, University of Pretoria. - 2008. - P. 34-39.

120. Mbati P.A., Hlatshwayo M., Mtshali M.S., Mogaswane K.R., de Waal T.D., Dipeolu O.O.. Ticks and tick-borne diseases of livestock belonging to resource - poor farmers in the eastern Free State of South Africa // Exp. Appl. Acarol. - 2002. - V. 28(1-4). - P. 217-224.

121. Miller R.J., Byford R.L., Smith G.S., Craig M.E., Vanleeuwen D. Influence of snakeweed foliage on engorgement, fecundity and attachment of the lone star tick (Acari: Ixodidae) // J. Agri. Agri. Entomol. - 1995. - V. 12(2/3). - P. 137-143.

122. Miller R.J., Davey R.B. and George J.E.. First report of organophosphate-resistant *Boophilus microplus* (Acari, Ixodidae) within the United States // Med. Entomol. - 2005. - V. 42. - P. 912-917.

123. Montasser A.A., Gadelhak G.G., Tariq S. Impact of ivermectin on the ultrastructure of the testis of *Argas (Persicargas) persicus* (Ixodoidea: Argasidae) // Exp. Appl. Acarol. - 2005. - V. 36. - P. 119-129.

124. Motoyashiki T., Tu A.T., Azimov D.A., Kazakov I. Isolation of anticoagulant from the venom of tick, *Boophilus calcaratus*, from Uzbekistan

// Thrombosis Research. - 2003. - V. 110. - P. 235-241.

125. Onofre S.B., Miniuk C.M., de Barros N.M., Azevedo J.L.. Pathogenicity of four strains of entomopathogenic fungi againest the bovine tick *Boophilus microplus* // Am. J. Vet. Res. - 2001. - V. 62. - P. 1478-1480.

126. Pandita N.N., Ram S. Control of ectoparasitic infestation in country goats // Small ruminant research. - 1990. - V. 3. - P. 403-412.

127. Pantaleoni R.A., Baratti M., Barraco L., Contini C., Cossu C.S., Filippelli M.T., Loru L., Romano M.. *Argas* (*Persicargas*) *persicus* (Oken, 1818) (Ixodida: Argasidae) in Sicily with considerations about its Italian and West-Mediterranean distribution // Parasite Dec. - 2010. - V. 17(4). - P. 349-55.

128. Paveglio S.A., Allard J., Mayette J., Whittaker L.A., Juncadella I. et al. The tick salivary protein, salp15, inhibits the development of experimental asthma // J. Immunol. Immunol. - 2007. - V. 178. - P. 7064-7071.

129. Pereira J.R., Famadas K.M., The efficiency of extracts *Dahlstedtia pentaphylla* (Leguminosae, Papilionoidae, Millettiedae) on *Boophilus microplus* (Canestrini, 1887) in artificially infested bovines // Vet. Paras. - 2006. - V. 142. - P. 192-195.

130. Permin A., Esmann J.B., Hoj C.H., Hove T. Ecto-, endo- and haemoparasites in free-range chickens in the Goromonzi District in Zimbabwe // Prevent. Veter. Med. - 2002. - V. 45. - P. 237-245.

131. Peter R.J., van den Bossche P., Penzhorn B.L., Sharp B. Tick, fly, and mosquito control-Lessons from the past, solutions for the future // Vet. Parasitol. - 2005. - V. 132(3-4). - P. 205-215.

132. Pirali-Kheirabadi K., Haddadzadeh H., Razzaghi-Abyaneh M., Zare R., Ranjbar-bahadori Sh., Nabian S., Rezaeian M. Preliminary study on virulence of some isolates of entomopathogenic fungi in different developmental stages of *Boophilus annulatus* in Iran. // J. Vet. Res. - 2007. - V. 62. - P. 113-118.

133. Pourseyed S.H., Tavassoli M., Bernousi I., Mardani K. Metarhizium

anisopliae (*Ascomycota: Hypocreales*): an effective alternative to chemical acaricides against different developmental stages of fowl tick *Argas persicus* (Akari: Argasidae) // Vet. Parasitol., 2010. Val. 20. №172 (3 -4). - P. 305 - 310.

134. Rajput Z.I., Song-hua Hu, Wan-jun Chen, Arijo A.G., Chen-wen Xiao. Importance of ticks and their chemical and immunological control in livestock // J. Zhejiang Univ. Sci B. - 2006. - V. 7(11). - P. 912-921.

135. Ramadan M.Y. Acaricidal and immunological studies on fowl tick *A. persicus* infecting commercial balady chickens flock // Third. Inter. Sci. Conf. - 2009. Benha & Ras Sudr, Egypt Fac. Vet Med. (Moshtohor), Benha Unuv. - P. 409-412.

136. Ramamoorthi N., Narasimhan S., Pal U., Bao F., Yang X.F. et al. The Lyme disease agent exploits a tick protein to infect the mammalian host // Nature. - 2005. - V. 436. - P. 573-577.

137. Regassa A. The use of herbal preparations for tick control in western Ethiopia // J. S. S. S. Afr. Vet. Assoc. - 2000. - V. 71(4). - P. 240-243.

138. Rezaie A.R. Kinetics of factor *Xa* inhibition by recombinant tick anticoagulant peptide: both active site and exosite interactions are required for a slow-and tight-binding inhibition mechanism // Biochem. - 2004. - V. 43. - P. 33683375.

139. Ribeiro J.M., Francischetti I.M. Role of arthropod saliva in blood feeding: Sialome and post-sialome perspectives // Ann. Rev. Entomol. - 2002. - V. 19. - P. 19.

140. Ribeiro J.M., Francischetti I.M. Role of arthropod saliva in blood feeding: sialome and postsialome perspectives // Annu Rev Entomol., 2003.-V. 48. - P. 7388.

141. Ribeiro V.L., Toigo E., Bordignon S.A., Goncalves K., von Poser G., Acaricidal properties of extracts from the aerial parts of *Hypericum polyanthemum* on the cattle tick *Boophilus microplus* // Vet. Parasitol. - 2007. - V. 147(1-2). - P. 199203.

142. Rolla G. Allergy to pigeon tick (*A. reflexus*): demonstration of specific

IgE binding components // Inter. Arch. Aller. Immunol. - 2004. - V. 135. - P. 293-295.

143. Salum M.R., Mtambuki A., Mulangila R.C. Designing a vaccination regime to control Newcastle disease in village chickens in the Southern zone of Tanzania // In: Proceedings of the joint 17th[th] Scientific Conf. Tanzania Soc. Animal Production and the 20th[th] Scientific Conf. Tanzania Veter. Assoc. Arusha (Tanzania). - 2002. - P. 299-305.

144. Samish M., Alekseev E.A. Arthropods as predators of ticks (Ixodoidea) // J. f Med. f Entomol. Entomol. - 2001. - V. 38. - P. 1-11.

145. Samish M., Ginsberg H., Glazer I. Biological control of ticks // Parasitol.- 2004. - V. 129. - P. 389 - 403.

146. Samish M., Glazer I. Entomopathogenic nematodes for the biocontrol of ticks // Trends in Parasitol. - 2001. - V. 17. - P. 368-371.

147. Sauer J.R., Essenberg R.C., Bowman A.S. Salivary glands in ixodid ticks: control and mechanism of secretion // J. Insect Physiol. Insect Physiol. - 2000. - V. 46. - P. 10691078.

148. Schwan T.G., Policastro P.F., Miller Z., Thompson R.L., Damrow T., Keirans J.E.. Tick-borne relapsing fever caused by *Borrelia hermsii*, Montana // EID Journal. - 2003. - V. 9(9).- P. 223-225/

149. Sewify G.H., Habib S.M. Biological control of the fowl tick *Argas persic A. persicus* by the entomopathogenic fungi *Beauvaria bassiana* and *Metarhizium anisopliae* // Sci. - 2001. - V. 74. - P. 121-123.

150. Shah A.H., Khan M.N., Iqbal Z., Safid M.S., Akhtar M.S. Some epidemiological aspects and vector role of tick infestation on layers in the Faisalabad district (Pakistan) // World's Poultry Scien. J. - 2006. - V. 62. - P. 145-157.

151. Shah A.H., Khan M.N., Iqbal Z., Sajid M.S. Tick Infestation in Poultry // Inter. J. Agriculture & Biology. - 2004. - V. 6. - P. 1162-1165.

152. Simons S.M., Junior P.L., Faria F., Batista I., Barros-Battesti D.M., Labruna M.B., Chudzinski-Tavassi A.M. The action of *Amblyomma*

cajennense tick saliva in compounds of the hemostatic system and cytotoxicity in tumor cell lines // Biomed. Pharmacother. - 2011. - V. 65(6). - P. 443-50.

153. Soltys J., Kusner L.L., Young A., Richmonds C., Hatala D., Gong B., Shanmugavel V., Kaminski H.J.. Novel complement inhibitor limits severity of experimentally myasthenia gravis // Ann. Neurol. - 2009. - V. 65(1). - P. 67-75.

154. Spiewak R., Lundberg M., Gunnar S., Johansson O., Buczek A. Allergy to pigeon tick (*A. reflexus*) in Upper Silesia, Poland // Ann. Agric. Environ. Med. - 2006. - V. 13(1). - P. 107-112.

155. Stoll P., Bassler N., Hagemeyer C.E., Eisenhardt S.U., Chen Y.C. et al. Targeting ligand-induced binding sites on GPIIb/IIIa via single-chain antibody allows effective anticoagulation without bleeding time prolongation // Arterioscler Thromb. Vasc. Biol. - 2007. - V. 27. - P. 1206-1212.

156. Tavassoli M., Ownag A., Meamari R., Rahmani S., Mardani K., Butt T.. Laboratory evaluation of three strains of the entomopathogenic fungus Metarhizium anisopliae for controlling *Hyalomma anatolicum anatolicum* and *Haemaphysalis punctata* // Int. J. Vet. Res. - 2009. - V. 3,1. - P. 11-15.

157. Telmadarehei Z., Vatandoust H., Rafinezhad J., Mohebali M., Tavakouli M., Abdi Goudarzi M., Faezeh F., Mandana A.A., Zareei Z.A., Jedari M., Mohtarami F., Azam Solgi A., Salarilak Sh., Mahdi E.R.. Frequency of Ixodidae and Argasidae ticks and evaluation of their sensitivity to sypermethrin in meshkinshahr // J. Ardabil univer. Ardabil univer. Med. Scien. (JAUMS). - 2009. - V. 9 (2 (32)). - P. 127-133.

158. Telmadarrehei Z., Nasirian H., Vatandoost H., Abuolhassani M., Tavakoli M., Zarei Z., Banafshi O., Rafinejad J., Salarielac S., Faghihi F.. Comparative Susceptibility of Cypermethrin in *Ornithodoros lahorensis* Neuman and *A. persicus Oken* (Acari: Argasidae) Field Populations // Pakistan J. Biol. Scien. - 2007. - V. 10. - P. 4315-4318.

159. Thembo M.K., Magano S.R., Shai L.J., The effects of aqueous root extract of Senna italica subsp. arachoides on the feeding performance of

Hyalomma marginatum rufipes adults // African J. Biotechnol. - 2010. - V. 9(7). - P. 10681073.

160. Tolleson D.R., Teel P.D., Stuth J.W., Strey O.F., Welsh T.H., Carstens G.E., Fecal NIRS: Detection of tick infestations in cattle and horses // Vet. Parasitol. - 2007. - V. 144. - P. 146-152.

161. Tomalski M.D., Kutney R., Bruce W.A., Brown M.R., Blum P.S., Travis I. Purification and haracterization of insect toxins derived from the mite Pyemotes tritici // Toxicon.-1989. - V. 27(10). - P. 1151-1167.

162. Tu A.T., Motoyashaki T., Azimov D.A. Bioactive compounds in tick and mite venoms (saliva) // Toxin Rev. - 2005. - V. 2. - P. 143-174.

163. Umrkulova S. Kh., Mirzayeva A. U., Akramova F. Ecological and faunistic researching of ticks of the Amblyomminae [Acari: Parasitiformes, Ixodidae] in Uzbekistan // The Eighth International Conference on European scientific development. Austria. Vienna, 2016. - V. 2. - P. 10 - 11.

164. Vandier J.Y. Le Guennec, Bedfer G. What are the signaling pathways used by norepinephrine to contract the artery? A demonstration using guinea pig aortic ring segments // Adv. Physiol. Educ. - 2002. - V. 26. - P. 195-203.

165. Viljoen G.I., Bezuidenhout J.D., Oberem R.T., Vermeulen N.J., Visser L., Gothet R., Netz A.W.. Izolation of a neurotoxin from the salivary glands of female Rhicephalus evertsi evertsi // J. Parasitol. Parasitol. - 1986. - V. 72(6). - P. 865-874.

166. Waxman L. Tick anticoagulant peptide (TAP) is a novel inhibitor of blood coagulation factor Xa // Science, 1990. - V. 248. - P.593-596.

167. Waxman L., Connolly T. M. Isolation of an inhibitor selective for collagen-stimulated platelet aggregation from the soft tick *Ornithodoros moubata* // J. Biol. Biol. Chem., 1993. - V. 268. - P. 5445-5449.

168. Williams L.D., Acaricidal activity of five marine algae extracts on female *Boophilus microplus* (Acari: Ixodidae) Florida // Entomol. - 1991. - V. 74(3). - P. 404-408.

169. Wutzler P., Sauerbrei A. Virucidal activity of the new disinfectant

monopercitic acid // Letters in Applied Microbiology. - 2004. - V. 39. - P. 194-198.

170. Zingali R.B., Jandrot-Perrus M., Guillin M.C., Bothrojaraein B.C., A new thrombin inhibitor isolated from *Bothrops jamraca* venom: characterization and mechanism of thrombin inhibition // Biochem. - 1993. - V. 10794-802.

Buy your books fast and straightforward online - at one of world's fastest growing online book stores! Environmentally sound due to Print-on-Demand technologies.

Buy your books online at
www.morebooks.shop

Kaufen Sie Ihre Bücher schnell und unkompliziert online – auf einer der am schnellsten wachsenden Buchhandelsplattformen weltweit! Dank Print-On-Demand umwelt- und ressourcenschonend produziert.

Bücher schneller online kaufen
www.morebooks.shop

info@omniscriptum.com
www.omniscriptum.com